R Programming and Its Applications in Water Resources Management

NIPA® GENX ELECTRONIC RESOURCES & SOLUTIONS P. LTD.
New Delhi-110 034

About the Editors

Dr. Naveena K, Scientist, LWMRG, CWRDM completed his PhD in Agricultural Statistics. He is a recipient of the Young Achiever Award, Society for Advancement of Human and Nature (2015), SP Dhall Distinguished Publication Award in Statistics Society for Advancement of Human and Nature (2015), and the Young Scientist Award ASTHA Foundation 2015. He is a member of the Indian Society of Agricultural Statistics (ISAS), Statistics Society for Advancement of Human and Nature (ASTHA) Foundation, and the Science and Tech Society for Integrated Rural Improvement. His research interests include Machine Learning Modeling, Forecasting Modeling, Multivariate Analysis, and Design of Experiment. He has more than 30 national and international publications.

Dr. K Ch V Naga Kumar is working as a 'Scientist' at Land and Water Management Research Group, KSCSTE-Centre for Water Resources Development and Management (CWRDM). He completed his PhD in Geo-Engineering (Remote Sensing and GIS) and M.Tech in Remote Sensing and GIS. He is a recipient of "Prof. K. Nageswara Rao Endowment Young Achiever Award-2016" given by the Indian Society of Geomatics (ISG) and International Union for Quaternary Research (INQUA) Travel grant under Young Researcher category for attending INQUA Congress held at Nagoya Japan during 26 July–2 August, 2015. He is a member of the Indian Society of Remote Sensing and the Indian Society of Geomatics. His research interest includes Coastal Morphodynamics and Holocene Delta Evolution. He has more than 40 publications and is handling several research projects pertaining to "Water Resources" funded from National & International agencies.

Dr. Surendran U, Principal Scientist, Division of Remote Sensing Application, ICAR-National Bureau of Soil Survey and Land Use Planning, Nagpur, completed his PhD and MSc in Soil Science and Agricultural Chemistry. He is a recipient of the Young Achiever award from SADHNA, Young scientist project by the Department of Science and Technology (DST-SERC-Fast track). He is the editorial board member of Agricultural Water Management, Elsevier publishers, handling editor of Scientific Reports, nature publishing group, and member of the International Society for Agricultural Meteorologists, Netherlands, etc. His research interests include Decision Support systems for soil nutrient budgeting and its impact on crop yield, Adsorption of nutrient/Modeling nutrient behavior in soils and its chemistry, Carbon Sequestration in soils, Climate change and its impact on agriculture, Crop Simulation Modeling, etc. He has more than 100 national and international peer reviewed publications. He handled research projects in Natural Resources Maangement funded by various National and International Agencies.

R Programming and Its Applications in Water Resources Management

Naveena K

Scientist B

Land and Water Management Research Group

KSCSTE-Centre for Water Resources Development and Management (CWRDM)

Kunnamangalam-673571, Kerala

K Ch V Naga Kumar

Scientist

Land and Water Management Research Group

KSCSTE-Centre for Water Resources Development and Management (CWRDM)

Kunnamangalam-673571, Kerala

Surendran U

Principal Scientist

Division of Remote Sensing Application

ICAR-National Bureau of Soil Survey and Land Use Planning

Nagpur-440033, Maharashtra

NIPA® GENX ELECTRONIC RESOURCES & SOLUTIONS P. LTD.

New Delhi-110 034

NIPA® GENX ELECTRONIC RESOURCES & SOLUTIONS P. LTD.

101,103, Vikas Surya Plaza, CU Block
L.S.C.Market, Pitam Pura, New Delhi-110 034
Ph : +91 11 27341616, 27341717, 27341718
E-mail:newindiapublishingagency@gmail.com
www: www.nipabooks.com

For customer assistance, please contact
Phone: + 91-11-27 34 17 17
Fax: + 91-11-27 34 16 16

Print ISBN: 978-93-58873-66-5

ebook SIBN: 978-93-58873-42-9

Composed and Designed by NIPA

Preface

Water is a valuable natural resource that is essential to all aspects of life on Mother "Earth" In fact, our planet is covered by 70% of water, but only 3% of the total water is in the form of freshwater, and again two-thirds of that is in the form of glaciers or is unavailable for our use. Presently, 1.1 billion people globally lack access to good drinking water resources, and another 2.7 billion find water scarce for at least one month of the year. While, India has an annual average precipitation of 1,170 mm and about 80% of the total area experiences an annual rainfall of 750 mm or more, which is due to large spatial and temporal variability in the rainfall. With growing urban population and land use changes, water scarcity will be exacerbated, and also with climate change and bio-energy demands amplify the complex relationship between world development and water demand. Another major threat in the recent past is climate change and its associated extreme floods and droughts events are a major concern, and there is a need to understand the changing dynamics in "Water Resources" with geographical space and time, for effective management of natural resources using the advanced tools like Remote Sensing and GIS technology, and also the use of R Programming.

Modern Remote Sensing (RS) and GIS technologies have revolutionized monitoring, analysis and modeling of the environments, resources and dynamics of the lithosphere, hydrosphere, atmosphere, and biosphere including anthropogenic issues and challenges that beset the present world. Significant research and development activities are being carried out in various scientific, academic, and industrial establishments relentlessly improving these technologies and their applications in every sphere of human activity and for societal benefit. Various applications have been developed, in the last three decades, wherein satellite remote sensing data is being put into use to provide quantitative and reliable information, thereby facilitating improved water resource management, for example Snow and Glaciers, Surface Water Resources, Irrigation Water Management, Reservoir Sedimentation, Hydro-Power generation, Interlinking of rivers, Flood Disaster monitoring and management, Watershed Management, River Engineering, Groundwater Prospecting, Environmental Impact Assessment and Management. And

"R Programming' is a well-developed, simple, and effective programming language and an integrated environment for statistical computing and data analysis. It provides a wide variety of statistical techniques (e.g., linear and nonlinear modeling, classical statistical tests, time-series analysis classification clustering, and raster data processing. The growth of R's usage in water science is reflected in the number of newly published packages connecting to water problems.

In this book, we explore the usage of different R packages to solve water science issues connecting to rainfall, drought, evapotranspiration, stream flow analysis, etc., advanced statistical and machine learning models like Modified Trend analysis, Autoregressive Integrated Moving Average (ARIMA) Artificial Neural Network (ANN), hybrid models (ARIMA-ANN, ARIMA-SVR) Multivariate modeling, spatiotemporal time series modeling, count time series modeling for forecasting, in addition to understanding the changing Water Resources.

Contents

Section 2: Development of Climate and Drought Indices

Section 3: Statistical Time Series Analysis for Water Resource Management

Section 5: Hybrid Time Series Modeling for Water Resource Management

Section 1

Basics of R Programming and Its Importance in Water Resource Analysis

1

An R-Based Approach Towards Understanding the Changing Dynamics in Water Resources

Naveena K[1], K Ch V Naga Kumar K.[1], Surendran U[1] Santhosh Onte[1], and Manoj P Samuel[1]

Centre for Water Resources and Development, Kunnamangalam, Kozhikode

Abstract

The R programming-based approach to water resource management combines various disciplines, including hydrology, climatology, ecology, geography, and social sciences, to develop a comprehensive understanding of water resources. This approach is crucial for addressing climate change impacts, promoting sustainable management of water resources, enhancing water security, addressing water quality issues, and promoting collaboration among stakeholders. It helps develop adaptation strategies to climate change impacts, identify alternative water sources, and promote water conservation practices. The approach involves using advanced analytical tools, such as statistical and machine learning techniques, to understand the complex processes affecting water resources. The recent development in R programming platforms is facilitating the implementation of the integrated approach more effectively. By combining the knowledge and expertise of different disciplines and using advanced analytical tools, we can effectively manage our precious water resources for future generations.

Keywords: *Ecology, water resources, R program, hydrology*

Introduction

An integrated approach to understanding the changing dynamics in water resources involves taking into account various factors that influence the management of water resources under changing climatic conditions

(Pahl-Wostl, 2007). It is a holistic approach that combines different disciplines, including hydrology, climatology, ecology, geography, and social sciences, to develop a comprehensive understanding of water resources. The integrated approach is essential for several reasons, including addressing climate change impacts, sustainable management of water resources, enhancing water security, addressing water quality issues, and promoting collaboration among stakeholders. Climate change is causing alterations in precipitation patterns, melting glaciers, and changes in river flows, affecting water availability and quality. An integrated approach helps in understanding the climate change impacts on water resources and developing adaptation strategies (Asif *et al.*, 2023).

Water resources are finite, and their management requires a comprehensive understanding of their interactions with the natural and social environment. An integrated approach helps develop sustainable management practices considering environmental, social, and economic factors. The increasing demand for water from population growth, urbanization, and industrialization has resulted in water scarcity in many parts of the world. An integrated approach can enhance water security by identifying alternative sources, developing water storage infrastructure, and promoting water conservation practices. Water quality is essential for human health and the environment. An integrated approach helps in identifying the sources of pollution, developing appropriate treatment technologies, and implementing measures to reduce pollution. Hydrology, climatology, and ecology are the key disciplines to understanding the physical processes of water resources. Geographic information systems (GIS) and remote sensing techniques are used to collect and analyze spatial data related to water resources. Social sciences, including economics and policy, are critical in understanding the social and economic impacts of water management decisions and in developing effective policies and regulations.

By joining the knowledge and expertise of various disciplines, the integrated approach to water resource management can help develop more effective and sustainable water management strategies. It enables us to identify potential threats and opportunities for water management, design appropriate interventions, and make informed decisions based on scientific evidence and stakeholder input. The integrated approach involves the use of advanced analytical tools, including statistical and modeling techniques, to understand the complex and interdependent processes affecting water resources. Statistical techniques can help detect trends in the data, while modeling techniques can simulate the behavior of hydrological systems and predict the impacts of different management strategies on water resources. R is a powerful open-source programming language and software environment that is widely used for statistical computing and data analysis. Recently, several R programming platforms have been developed to facilitate the implementation of the integrated water management approach more effectively.

In conclusion, the integrated approach to understanding the changing dynamics in water resources is essential for developing sustainable water management practices, addressing climate change impacts, enhancing water security, addressing water quality issues, and promoting collaboration among stakeholders. By combining the knowledge and expertise of different disciplines and using advanced analytical tools, we can effectively manage our precious water resources for future generations.

R for Understanding the Changing Dynamics in Water Resources

R programming is open-source software that allows for flexible and customizable data analysis and visualization. It is particularly useful for managing and analyzing large datasets related to water resources, which can include data on precipitation, river flows, water quality, and climate forecasts (Zinck, 2018). R programming is a powerful tool for developing complex statistical models that can be used to forecast and predict future trends related to water resources. For instance, time series analysis in R can be used to model and predict water availability based on historical data and climate projections. Moreover, R offers a wide range of machine learning algorithms that can help identify patterns and relationships in large datasets, allowing for more accurate predictions and informed decision-making. These features make R an invaluable tool for water resource management and planning, helping to confirm the sustainable use of this precious natural resource. In addition to its data analysis capabilities, R also offers powerful data visualization tools, which can be extremely useful in presenting water resource data in a clear and informative manner. By creating visual representations of data, users can better understand patterns and trends, identify anomalies, and make more informed decisions. R provides several packages for creating highly effective visualizations of water resource data, including ggplot2 for static graphics, plotly for interactive charts, leaflet for interactive maps, rgl for 3D graphics, and dygraphs for time series charts. These packages offer users an extensive range of options for exploring and presenting data, making it easier to analyze complex water resource issues and develop more effective management strategies.

R programming can be a valuable tool for hydrological modeling, allowing for the analysis and modeling of water flows, precipitation, and temperature data to better understand water resource dynamics. In addition, R programming provides several packages that can assist in hydrological modeling, such as hydroTSM for time series analysis, SWAT model for simulating land use impacts on water resources, ecohydro for ecohydrological modeling, water balance for calculating water balance components, and hydroGOF for evaluating model performance. By leveraging these packages, users can gain insight into hydrological data and make informed water management decisions while developing strategies for sustainable use of water resources.

R is a powerful tool for developing indices related to water resources, such as drought indices, evapotranspiration indices, and extreme weather indices. These indices can help identify regions more prone to water scarcity and aid in developing targeted water management strategies. To achieve this, R provides several packages for developing water resource indices, such as SPEI, ET, climdex. pcic, TSEB, and hydrostats (Slater *et al*, 2019; Astagneau *et al*., 2020). These packages offer functions for calculating various indices related to water resources, such as drought indices, evapotranspiration indices, extreme weather indices, and hydrological statistics. These indices can be used to evaluate water availability and use in a watershed, identify areas that are experiencing water scarcity, and develop effective water management strategies to ensure the sustainable use of water resources.

Overall, the use of R programming in water resource management can enhance our understanding of the changing dynamics in water resources by providing advanced statistical models and effective data visualization tools. This, in turn, can help us to make informed decisions about water resource management and sustainability.

References

Asif, Z., Chen, Z., Sadiq, R., Zhu, Y. "Climate Change Impacts on Water Resources and Sustainable Water Management Strategies in North America." Water Resources Management, (2023): 1–16.

Astagneau, P.C., Guillaume, T., Olivier, D., Guillaume, J. H., Juraj, P., Brauer, et al. "Hydrology Modelling R packages: A Unified Analysis of Models and Practicalities from a User Perspective." Hydrology and Earth System Sciences Discussions, 2020.

Pahl-Wostl, C. "Transitions towards adaptive management of water facing climate and global change." Water Resources Management, 21, no. 1 (2007): 49–62.

Slater, L.J., Thirel, G., Harrigan, S., Delaigue, O., Hurley, A., Khouakhi, A., et al. "Using R in Hydrology: A Review of Recent Developments and Future Directions." Hydrology and Earth System Sciences, 23, no. 7 (2019): 2939–63.

Zinck, N. 2018. Development of Application Software for Water System Data Management, Visualization, and Analysis With the Shiny Framework.

2

An Introduction to R Programming for Water Resource Management

Naveena K[1], Santosha Rathod[2], and K Ch V Naga Kumar[1]

[1]*Centre for Water Resources Development and Management, Kunnamangalam Kozhikode*

[2]*ICAR-Indian Institute of Rice Research, Hyderabad*

Abstract

R programming language and R Studio have revolutionized the way water resource management is conducted, allowing hydrologists to perform complex statistical computations and data analysis, which are essential for effective water resource management. R's powerful graphics capabilities enable the visualization of water resource data, making it easier to understand and present results to stakeholders. In water resource management, R is used to analyze water availability, predict future water demand, and assess the impact of climate change on water resources. R packages designed for hydrology enable the creation of models for water resource management, letting hydrologists predict and manage water resources more effectively. The open-source nature of R and its compatibility with other programming languages like C, C++, Java, and Python makes it an ideal tool for integrating data from multiple sources and conducting complex analyses. Additionally, the availability of online resources, including tutorials, forums, and packages, makes it easy for water resource managers to learn and apply R to their work. Overall, R programming has become a vital tool for water resource management, enabling hydrologists to effectively manage water resources and plan for future water needs.

Keywords: *Hydrology, statistical analysis, R studio, R program, water resource management*

Introduction

R language is a popular and versatile tool for hydrological data analysis and visualization, offering several powerful packages for handling and analyzing hydrological parameters such as river flow, precipitation, groundwater levels, and water quality parameters. R language provides a wide range of statistical functions for hydrological parameter analysis, including time-series analysis, frequency analysis, trend analysis, and regression analysis. Moreover, R's powerful graphics capabilities make it possible to create high-quality and customizable visualizations of hydrological data. In addition to its statistical and graphical capabilities, R also offers a wide range of packages for data manipulation and cleaning, making it a powerful tool for preprocessing hydrological data. R's ability to handle and analyze large datasets also makes it well-suited for processing and analyzing large hydrological datasets. Furthermore, R has a strong community of developers and users who regularly contribute new packages and tools, providing a constantly evolving stage for hydrological analysis. Overall, R is a versatile and powerful tool for hydrological data analysis and visualization, offering numerous capabilities and resources for informed decision-making in water resources management and related fields.

About R

R is a programming language for statistical data analysis, developed as an extended version of the S language. It was created by "Ross Ihaka and Robert Gentlemen" at the University of Auckland in 1993 and is currently maintained by the R Core Team. The name "R" comes from the first names of the language's makers and its inheritance from the S language. R is widely used among statisticians and data miners for developing statistical software and data analysis. It is free, open source, and capable of almost any statistical analysis, including the most recently developed statistical methodologies. R is easily extensible through new contributions from statisticians and researchers in the form of packages, which can be found on the Comprehensive R Archive Network (CRAN). R has a wide range of statistical functions for data analysis, including time-series analysis, frequency analysis, trend analysis, and regression analysis. It is commonly used for hydrological data analysis and visualization, as it offers several powerful packages for handling and analyzing hydrological parameters such as river flow, precipitation, groundwater levels, and water quality parameters. R can be used on various operating systems such as Windows, Linux, and macOS. Separate versions and packages are available for each operating system. The current version of R for Windows OS is 4.1.2.

Getting and Installing R

R is a free software that can be downloaded from CRAN at http://cran.r-project.org/. It is recommended to choose a mirror site that is closest to your location for faster download.

To install R on your computer, follow these steps:

- Double-click on the downloaded R.exe file.
- Select "English" as the language for the installation.
- The R setup wizard window (Fig. 1) will appear.
- Click on "Next" and accept the default settings during the installation.

As of April 1, 2023, the latest version of R for Windows is R 4.1.3.

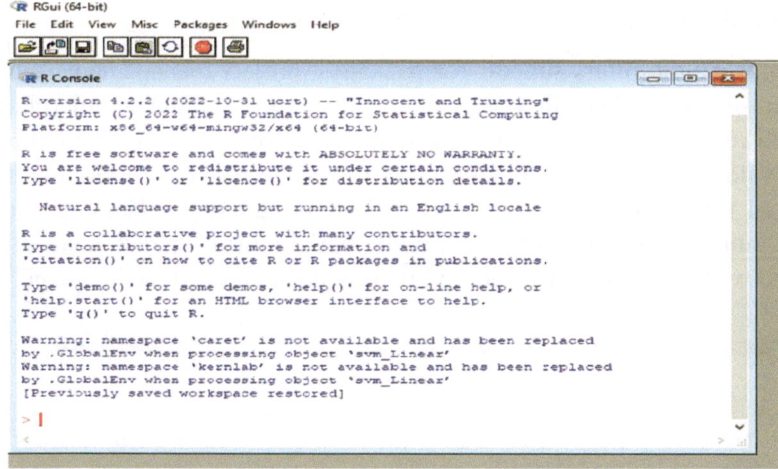

Fig. 1: R console window.

Starting R

To start R, simply click on the R icon on your desktop or go to the start menu, select all programs, then click on R, and select the latest version installed. Once launched, the R Console screen will appear, where all R code is written and outputs are displayed. In the R Console screen, you will find a toolbar with several buttons and a few menus in the R window. To learn what each button on the toolbar does, just hover your mouse over it and a description of the button will appear. There is also an R editor window that can be used to write and edit R code. The R editor window is similar to a text editor, with features like select, cut, copy, paste, typing text, and deleting text. To open the

R editor window, click on File and then select New Script from the menu bar. However, many users prefer to use R Studio, a free and open-source integrated development environment (IDE) for R. R Studio provides an improved user interface, as well as many additional features and capabilities that make working with R easier and more efficient.

R Studio

R Studio (Fig. 2) is an IDE that is specifically designed to work with the R programming language. It is available in two different editions: R Studio Desktop and R Studio Server. The latest version of R Studio Desktop as of April 2023 is R Studio v1.4.1106. It is intended to be used as a regular desktop application and is available for Mac OS X, Windows, and Linux. It provides a user-friendly interface for working with R, which includes a console for entering R code, a syntax-highlighting editor with direct code execution capabilities, and tools for plotting, debugging history, and workspace management.

In addition, R Studio offers support for packages, project management, and version control systems like Git. It is available in open-source and commercial editions and runs on various operating systems. The free edition of R Studio can be downloaded from the official R Studio website. Overall, R Studio is a useful tool for anyone who works with the R programming language. It provides a more convenient and user-friendly environment for data analysts who need to work with large datasets and those who are new to the language.

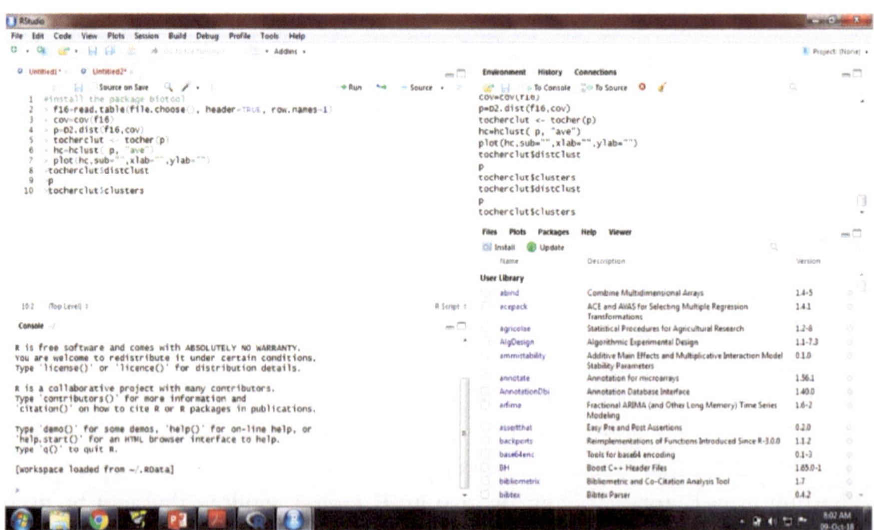

Fig. 2: R Studio.

The program has a user-friendly interface that consists of four main windows which are as follows:

1. The top left window is the editor window where you can write and edit R code. The editor window has many useful features such as syntax highlighting, auto-completion, and code snippets that make writing R code easier and faster.
2. The bottom left window is the console window where R code is executed. The console window is similar to the R console but it has some additional features like code completion, syntax highlighting, and error highlighting.
3. The top right window of R Studio has two tabs: Environments and History. In the Environments tab, you can see which objects are currently loaded in R, and you can also create and manage your own environments. The History tab provides a history of commands that have already been executed.
4. The bottom right window of R Studio has several tabs: Files, Plots, Packages, Help, and Viewer. The Files tab can be used to explore and navigate different files and directories. The Plots tab is used to view plots produced from R code. The Packages tab shows the packages that are already installed and also allows easy installation and loading of new packages in a session. The Help tab can be used to search for help on R functions and packages, and the Viewer tab can be used to view local web content.

R Commands and Common Rules

R commands are entered in the Console, which is the primary interface for working with R. The Console is where you type commands to tell R what to do. The > symbol in the Console specifies that R is ready to receive a command. R is an interactive language, which means that it responds to commands in real-time. When you type a command in the Console and press Enter, R reads the command and returns the results or an error message if there is a syntax or other error. R has a syntax that must be followed when typing commands. Commands are case sensitive, which means that AB, Ab, aB, and ab are all different objects in R. R commands can include alphanumeric characters, periods, and spaces. A period is often used to separate object names from their attributes, such as y.mean OR x.length. Comments can be added to the R code using the # symbol. Anything after the # symbol is ignored by R and is not evaluated as part of the code.

The working directory is the folder where R is currently working. By default, the working directory is set in the user's documents folder. However, it can be changed using the setwd() function. You can customize the getwd() function to find out the current working directory. R can read and write files in the working directory directly, which makes it convenient to work with data files and code files in the same folder. It is recommended to set the working directory to the folder where most data files and code files are located. This helps to avoid errors when reading or writing files and makes it easier to work with files in a consistent manner. By setting the working directory at the beginning of an R session, you can avoid having to type the full file path for every file that you want to read or write.

R Packages for Hydrology

There are many R packages available for hydrology, each with its own set of functions and capabilities (Slater *et al.*, 2019; Astagneau *et al.*, 2020). These packages can be used to analyze time series data, calculate hydrological indices, model hydrological processes, and generate synthetic hydrological data, among other tasks.

Some of the most commonly used R packages for hydrology include:

- hydroTSM: This package provides functions for time series analysis and visualization of hydrological data, including tools for missing data imputation, trend, and extreme value analysis.
- SWAT model: This package provides an interface to the Soil and Water Assessment Tool, a widely used model for simulating water quality and quantity in watersheds.
- hydromad: This package provides functions for fitting and evaluating the performance of hydrological models using a variety of statistical techniques.
- SPEI: This package provides functions for calculating the Standardized Precipitation-evapotranspiration Index, a drought index that considers both precipitation and temperature.
- hydroGOF: This package provides functions for evaluating the goodness of fit of hydrological models, including measures of model efficiency, bias analysis, and correlation.
- streamMetabolism: This package provides functions for estimating stream metabolism rates, including gross primary production and ecosystem respiration, from a chronological series of dissolved oxygen data.

- evapotranspiration: This package provides functions for calculating evapotranspiration using several different methods, including the Penman-monteith method, the Priestley-taylor method, and the Hargreaves-samani method.

  To use these packages, you first need to install them in your R environment using the **install.packages()** function. Once installed, you can load the package using the **library()** function, and then access its functions and capabilities. Many packages also come with documentation and example code to help you get started.

  Overall, R packages are a potent and flexible tool for hydrologists, providing a wide range of functions and capabilities for analyzing and modeling hydrological data.

References

An Introduction to R Notes on R: A Programming Environment for Data Analysis and Graphics. Version 4.1.2 (2022-11-01)

Astagneau, P. C., Guillaume, T., Olivier, D., Guillaume, J. H., Juraj, P., Brauer, C.C., et al. (2020). Hydrology modelling R packages: A Unified Analysis of Models And Practicalities from a User Perspective. Hydrology and Earth System Sciences Discussions, (2020).

Slater, L.J., Thirel, G., Harrigan, S., Delaigue, O., Hurley, A., Khouakhi, et al. "Using R in Hydrology: A Review of Recent Developments and Future Directions." Hydrology and Earth System Sciences 23, no. 7 (2019): 2939–63.

3

Data Visualization Tools and Techniques for Analyzing Hydrological Data in R

Naveena K and K Ch V Naga Kumar

Centre for Water Resources and Development, Kunnamangalam, Kozhikode

Abstract

Hydrological data visualization is essential for gaining insight into the behavior and characteristics of water resources. R Language offers numerous powerful packages for data visualization, each with its own strengths and limitations. One popular package for visualizing hydrological data is hydroTSM, which provides functions for reading, deploying, and plotting time-series data. With hydroTSM, we can create time-series plots of variables, such as precipitation, river flow, and temperature, with options for adding trend lines, error bars, and annotations. Another package commonly used for hydrological data visualization is plotly, which offers interactive and dynamic plots that can be easily shared and embedded in web applications. Plotly provides a range of plot types, including scatterplots, line plots, and heatmaps, which can be customized with options for color scales, annotations, and tooltips. In addition to these packages, the ggplot2 package offers a dominant tool for creating publication-quality visualizations of hydrological data. With ggplot2, we can use the grammar of graphics to create a wide range of plot types including time-series plots, spatial maps, and scatterplots. ggplot2 also provides options for customizing plot elements such as axis labels, legends, and plot themes. Overall, each of these packages offers unique strengths and can be used to create clear and meaningful visualizations of hydrological data. By choosing the appropriate package and plot type for a given dataset, we can gain insight into the patterns and trends in the data, which can inform decision-making in fields such as water resources management and climate adaptation.

Keywords*: *Data visualization, ggplot2, hydroTSM, plotly, R-studio, R program

Introduction

Hydrological data visualization plays a crucial role in analyzing and understanding the behavior and characteristics of water resources. By visualizing hydrological data, we can gain insight into the distribution, variability, and trends of different hydrological variables such as river flow, precipitation, groundwater levels, and water quality parameters. This information can help decision-making in areas such as water resources management, flood forecasting, and climate adaptation.

R Language is a popular and versatile tool for hydrological data visualization, offering several powerful packages for data analysis and visualization. Each of these packages has its own strengths and limitations, which can be leveraged depending on the specific needs of a given analysis. For example, ggplot2 is an extensively used package for creating publication-quality static plots that are customizable and visually appealing. This package is based on the grammar of graphics, which provides a new plotting system that combines the benefits of both base and lattice graphics. ggplot2 can be used to create a wide range of plot types, including time-series plots, spatial maps, and scatterplots. It provides several options for customizing plot elements such as axis labels, legends, and plot themes. On the other hand plotly offers interactive and dynamic plots that can be easily shared and embedded in web applications. With plotly, we can create plots that allow for zooming, panning, and hovering over data points to display additional information. This package is mainly useful for exploring relationships between hydrological variables. In addition to these packages, hydroTSM provides functions for reading, manipulating, and plotting time-series data. hydroTSM can help us create time-series plots of variables such as river flow, precipitation, and temperature, with options for adding trend lines error bars, and annotations.

Overall, R Language offers several powerful packages for hydrological data visualization, each with its own strengths and limitations. By selecting the appropriate package and plot type for a given dataset, we can gain insight into the patterns and trends in the data, which can help decision-making in the management and conservation of water resources.

The HydroTSM Package

The hydroTSM package in R is specifically designed for time-series analysis and provides functions for reading, manipulating, and plotting hydrological time-series data (Zambrano-Bigiarini, 2017). The package supports different data formats such as Excel, CSV, and NetCDF, which can be read using the read.zoo() function.

To create time-series plots using hydroTSM, we typically start by creating a time-series object using the zoo() function. This object stores the time-series data along with metadata such as start and end dates, frequency, and time zone. Once we have a time-series object, we can create a plot using the plot() function. hydroTSM provides numerous options for customizing the plot, including:

- Adding trend lines using the **lines()** function.
- Adding error bars using the **arrows()** function.
- Adding annotations such as titles, labels, and legends using the **title()**, **xlab()**, **ylab()**, and **legend()** functions.
- Adjusting plot parameters such as axis limits, tick marks, and font size using the **par()** function.

In addition to basic time-series plots, hydroTSM also supports more advanced visualizations such as stacked plots, box plots, and scatter plots. These plot types can be useful for comparing several time-series data sets or exploring relationships between variables. Overall, hydroTSM provides a powerful tool for analyzing and visualizing hydrological time-series data in R. With its functions for reading, manipulating, and plotting time-series data, hydroTSM can help hydrologists and water resources managers gain insight into the behavior and trends of water resources over time.

llustration of Different Rainfall Plots Using the hydroTSM

In this study, we analyzed 33 years of daily rainfall data (1/1/1986 to 12/31/ 2020) from the Vythiri station in Wayanad, India. We utilized the hydroplot function from the hydroTSM package to explore and visualize the time series data efficiently.

Hydroplot of Daily Rainfall

The hydroplot() function creates multi-panel plots including time series plots, boxplots, and histograms (Fig. 1). This allowed us to gain insight into the distribution and patterns of the rainfall data over time, and identify any outliers or trends that may be present.

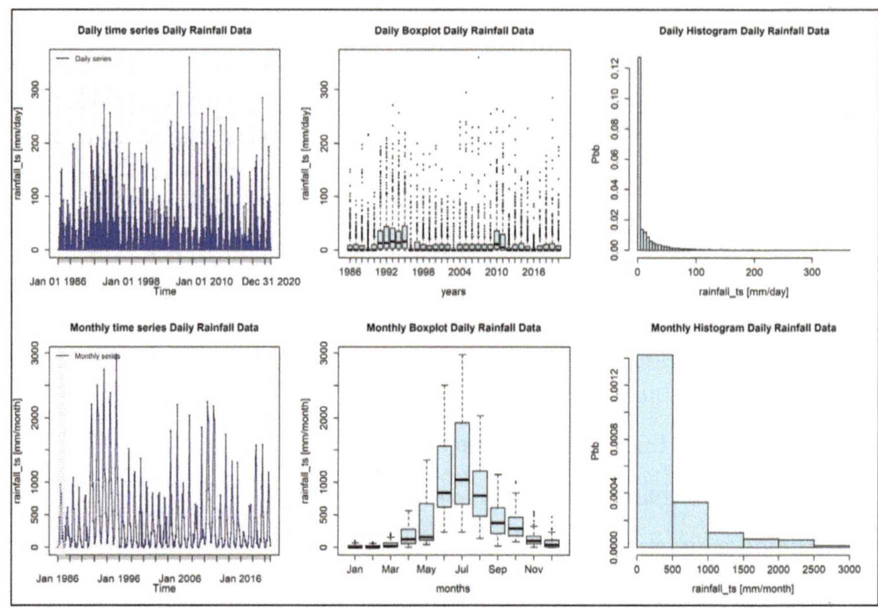

Fig. 1: Time series plots, boxplots, and histograms.

```
## r code for hydroplot of daily rainfall data
#Load libraries after installation of packages
library(hydroTSM)
library(ggplot2)
# Define start and end dates of the series
start_date <- as.Date("1986-01-01")
end_date <- as.Date("2020-12-31")
# Create date sequence
date_seq <- seq(from=start_date, to=end_date, by="day")
# Create empty data frame with date column
rainfall_data1 <- data.frame(date=date_seq)
# Import rainfall data from file (assuming the file has one column of precipitation data)
rainfall_data <- read.table(file="clipboard", header = TRUE)
###Add this data to Created empty data frame with date column
rainfall_data1$rainfall<-rainfall_data$rainfall
# Convert to time-series object
rainfall_ts <- zoo(rainfall_data1$rainfall, order.by=rainfall_data1$date)
# Set the plotting device to JPEG format
jpeg("my_plot.jpg", width = 12, height = 8, units = "in", res = 400)
# Plot the precipitation data
hydroplot(rainfall_ts, var.type="Precipitation", main="Daily Rainfall Data",
      pfreq = "dm", from="1986-01-01", to="2020-12-31")
# Turn off the plotting device
dev.off()
```

Visualizing the Wet and Dry Seasons—Monthly Rainfall

The matrix plot is a type of heatmap (Fig. 2) that displays the monthly precipitation values as a grid of colored rectangles, with dimmer colors representing higher precipitation values and lighter colors representing lower precipitation values. The x-axis shows the months of the year and the y-axis the years in the dataset.

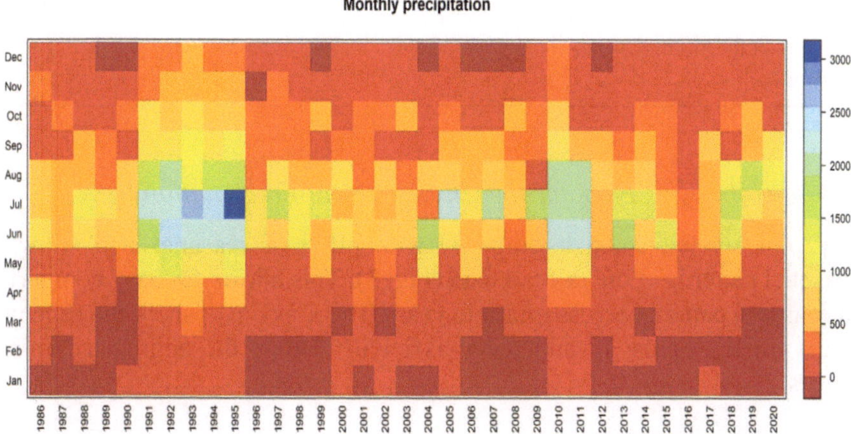

Fig. 2: The matrix plot.

R code for heatmap

```
###Load required packages
library(hydroTSM)
library(lattice)
# Define start and end dates of the series
start_date <- as.Date("1986-01-01")
end_date <- as.Date("2020-12-31")
# Create date sequence
date_seq <- seq(from=start_date, to=end_date, by="day")
# Create empty data frame with date column
rainfall_data1 <- data.frame(date=date_seq)
# Import rainfall data from file (assuming the file has one column of rainfall data)
rainfall_data <- read.table(file="clipboard", header = TRUE)
###add this data to Created empty data frame with date column
rainfall_data1$rainfall<-rainfall_data$rainfall
# Convert to time-series object
rainfall_ts <- zoo(rainfall_data1$rainfall, order.by=rainfall_data1$date)
# Daily zoo to monthly zoo
m <- daily2monthly(rainfall_ts, FUN=sum, na.rm=TRUE)
# Creating a matrix with monthly values per year in each column
```

```
M <- matrix(m, ncol=12, byrow=TRUE)
colnames(M) <- month.abb
rownames(M) <- unique(format(time(m), "%Y"))
# Plotting the monthly precipitation values
# Set the plotting device to JPEG format
jpeg("my_plot1.jpg", width = 12, height = 12, units = "in", res = 400)
# Plot the rainfall data
print(matrixplot(M, ColorRamp="Precipitation",
        main="Monthly precipitation"))
# Turn off the plotting device
dev.off()
```

Seasonal Analysis Plots for the Chronological Precipitation Series

The plot created using hydroplot() with pfreq of seasonal will show the seasonal patterns in the rainfall data (Fig. 3). This allows for easy comparison of seasonal patterns across multiple years. The x-axis will show the time units (e.g., months or days), and the y-axis will display the aggregated rainfall values.

```
###seasonal plot
###Load packages
library(hydroTSM)
# Define start and end dates
start_date <- as.Date("1986-01-01")
end_date <- as.Date("2020-12-31")
# Create date sequence
date_seq <- seq(from=start_date, to=end_date, by="day")
# Create empty data frame with date column
rainfall_data1 <- data.frame(date=date_seq)
# Import rainfall data from file (assuming the file has one column of rainfall data)
rainfall_data <- read.table(file="clipboard", header = TRUE)
###add this data to Created empty data frame with date column
rainfall_data1$rainfall<-rainfall_data$rainfall
# Convert to time-series object
rainfall_ts <- zoo(rainfall_data1$rainfall, order.by=rainfall_data1$date)
# Set the plotting device to JPEG format
jpeg("my_plot3.jpg", width = 12, height = 12, units = "in", res = 400)
hydroplot(rainfall_ts, pfreq="seasonal", FUN=sum, stype="default")
# Turn off the plotting device
dev.off()
```

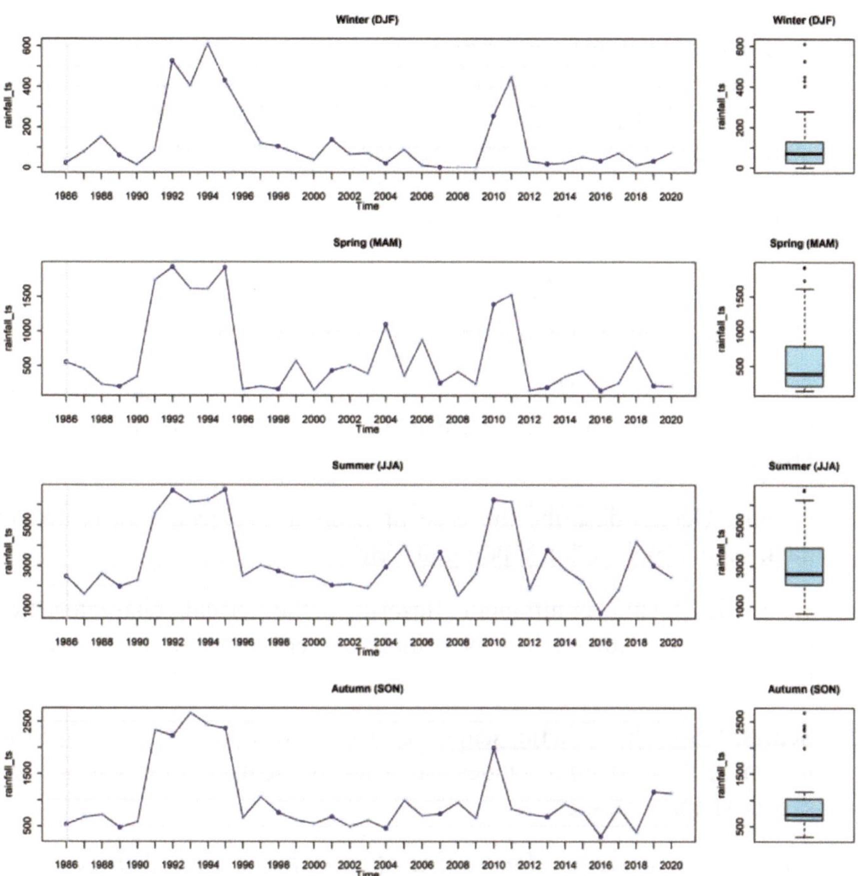

Fig. 3: Seasonal patterns in the rainfall data.

The ggplot2 Package

ggplot2 is a powerful R package for making visualizations that can be useful for analyzing hydrological variables. One common use of ggplot2 for hydrological data visualization is to create time series plots. Time series plots can show the variation of a hydrological variable over time, such as the changes in precipitation over a year or the seasonal trends in streamflow, water quality, temperature, etc. Another useful feature of ggplot2 for hydrological data visualization is its ability to create maps. Hydrological variables often vary spatially such as changes in precipitation or groundwater levels across a watershed. ggplot2 can be used to make maps that display these spatial variations. The maps can be customized to include different layers, such as hydrological features (such as streams or watersheds), topography, or land use. By using ggplot2 for hydrological mapping, it is possible to easily create publication-quality maps that display complex spatial information in a clear and visually appealing way. Overall, ggplot2 is a potent tool for visualizing hydrological data. It provides a flexible and customizable framework for displaying complex information in a clear and visually appealing way.

The components of ggplot2 are as follows:

- Data: Data must be stored as an R data frame in ggplot2.
- Coordinate system: This component describes the 2-D space that data is projected onto, such as Cartesian coordinates, polar coordinates, map projections, etc.
- Geoms: Geoms describe the type of geometric objects that represent data, such as points, lines, polygons, etc.
- Aesthetics: This component describes the visual characteristics that represent data, such as position, size, color, shape, transparency, fill, etc.
- Scales: For each aesthetic, scales describe how the visual characteristic is converted to display values, such as log scales, color scales, size scales, shape scales, etc.
- Stats: Stats describe the statistical transformations that typically summarize data, such as counts, means, medians, regression lines, etc.
- Facets: Facets describe how the data is split into subsets and displayed as multiple small graphs.

By utilizing these components, ggplot2 can create attractive, hassle-free plots that take care of minor details like drawing and displaying legends, allowing you to focus on analyzing and interpreting the data. Even experienced R users find ggplot2 to be a valuable tool for making graphics in R, and its deep grammar of graphics provides flexibility and customization for creating a wide variety of visualizations.

Illustration of rainfall plot using simulated monthly rainfall (Fig. 4) with the inclusion of all components of ggplot2

```
# First, generate some example data for a simulated rainfall time series
set.seed(123)
n <- 100 # number of time points
rainfall <- rnorm(n, mean = 50, sd = 10) # simulated rainfall data
# Load ggplot2 library
library(ggplot2)
# Create a ggplot object with the rainfall data as the data frame
p <- ggplot(data = data.frame(x = 1:n, y = rainfall), aes(x = x, y = y))
# Add a line geom to display the rainfall time series
p <- p + geom_line(color = "steelblue", size = 1.5)
# Add a horizontal line to indicate the average rainfall
avg_rainfall <- mean(rainfall)
p <- p + geom_hline(yintercept = avg_rainfall, color = "darkgreen", linetype = "dashed")
# Add a shaded area to represent the range of normal rainfall values
lower <- quantile(rainfall, 0.1)
upper <- quantile(rainfall, 0.9)
p <- p + geom_ribbon(aes(ymin = lower, ymax = upper), alpha = 0.2, fill = "gray")
# Add labels for the x and y axes
p <- p + labs(x = "Time", y = "Rainfall (mm)")
# Add a title for the plot
p <- p + ggtitle("Simulated Rainfall Time Series")
# Add a caption for the plot
p <- p + labs(caption = "Simulated rainfall data from a hydrological model")
# Use a different theme to change the appearance of the plot
library(ggthemes)
p <- p + theme_few()
# Display the plot
p
```

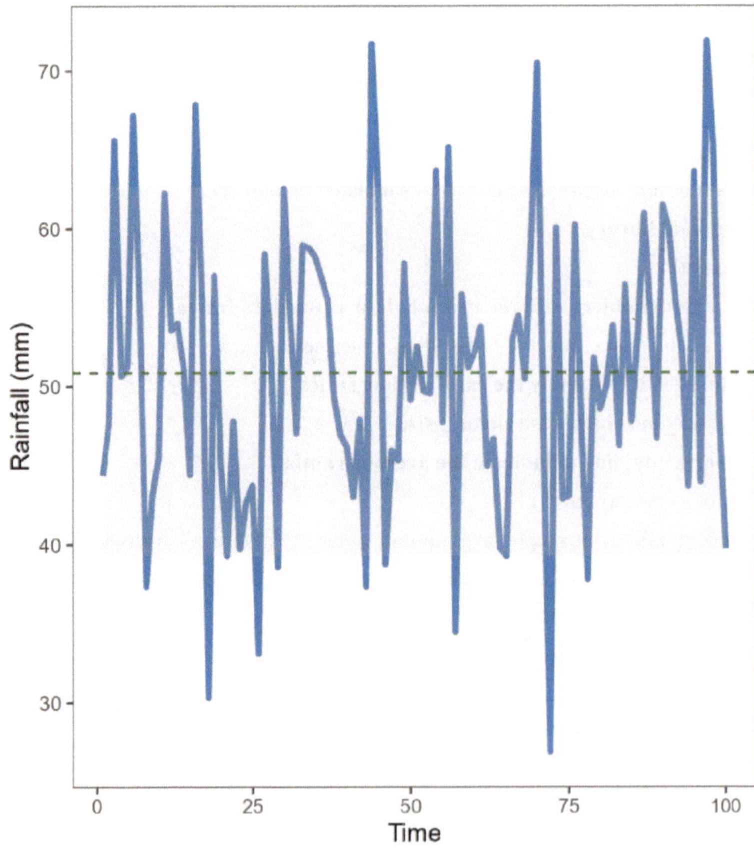

Fig 4: Rainfall plot using simulated monthly rainfall.

Illustration of rainfall plot by means of simulated monthly rainfall from two stations (Fig. 5) with inclusion of all components of ggplot2

```
# Example data for two stations
set.seed(123)
n <- 100 # number of time points
rainfall_station1 <- rnorm(n, mean = 50, sd = 10) # simulated rainfall data for station 1
rainfall_station2 <- rnorm(n, mean = 70, sd = 15) # simulated rainfall data for station 2
# Create a data frame with the rainfall data for both stations
df <- data.frame(x = 1:n, y1 = rainfall_station1, y2 = rainfall_station2)
# Load ggplot2 library
library(ggplot2)
```

```
# Create a ggplot object with the rainfall data as the data frame
p <- ggplot(data = df, aes(x = x))
# Add a line geom to display the rainfall time series for station 1
p <- p + geom_line(aes(y = y1), color = "steelblue", size = 1.5)
# Add a line geom to display the rainfall time series for station 2
p <- p + geom_line(aes(y = y2), color = "darkred", size = 1.5)
# Add labels for the x and y axes
p <- p + labs(x = "Time", y = "Rainfall (mm)")
# Add a title for the plot
p <- p + ggtitle("Simulated Rainfall Time Series for Two Stations")
# Add a caption for the plot
p <- p + labs(caption = "Simulated rainfall data from a hydrological model")
# Use a different theme to change the appearance of the plot
library(ggthemes)
p <- p + theme_few()
# Display the plot
p
```

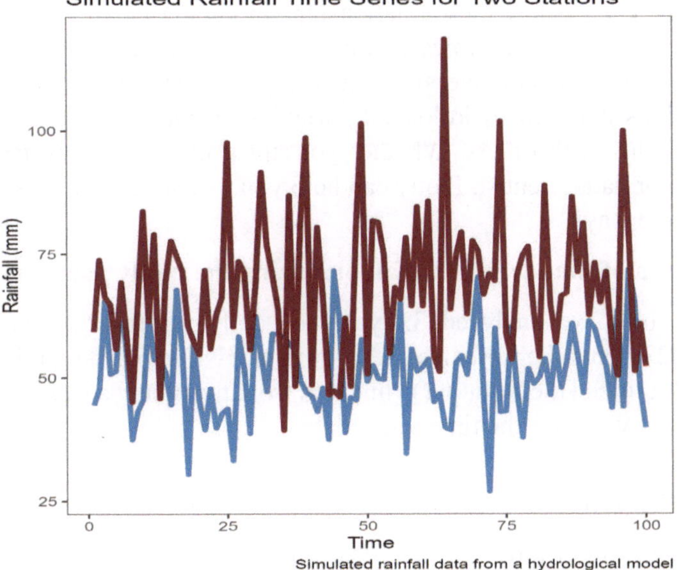

Fig. 5: Rainfall plot by means of simulated monthly rainfall from two stations.

Plotly Package of R

Plotly is a powerful data visualization library that allows users to create interactive plots that can be easily shared and embedded in web applications. With Plotly, users can create a wide range of plots, including scatter plots, line plots, heatmaps, bar charts, and more. These plots can be customized with a variety of options, including color schemes, markers, and line styles.

One of the key features of Plotly is its ability to create interactive plots. These plots allow users to zoom in and out, pan across the plot, and hover over data points to display additional information. For example, if you have a scatter plot showing the relationship between rainfall and river flow, you can hover over a data point to see the exact values of rainfall and flow for that point.

Another useful feature of Plotly is its ability to display data in real time. For example, if you have a sensor network collecting data on water quality, you can use Plotly to create a real-time plot that updates automatically as new data is available. This can be a powerful tool for the assessment of water quality and detecting changes in the system over time.

Plotly is also useful for exploring relationships between hydrological variables. For example, you can use Plotly to create a scatter plot showing the relationship between precipitation and streamflow, or a heatmap showing the spatial distribution of groundwater levels. By creating interactive plots with Plotly, one can explore these relationships in greater detail and gain insights into the underlying dynamics of the system. Overall, Plotly is a powerful tool for creating interactive and dynamic plots that can be used to explore relationships between hydrological variables, display data in real-time, and share insights with others. Whether you are a hydrologist, water resources manager, or data scientist, Plotly can help you visualize your data in new and informative ways.

Illustration of time series rainfall plot using the plotly package.

In this study, we analyzed 33 years of daily rainfall data (1/1/1986 to 12/31/2023) from the Ambalvayal, Vythiri, and Nilambur station in Wayanad, India. We utilized the "plot_ly()" function from the ploty package to efficiently explore and visualize the time series data.

Annual Rainfall Plot Creation

An annual rainfall plot is a graphical representation of the total amount of rainfall received in a given region over a year. This plot typically shows the variation of monthly rainfall amounts and the total annual rainfall in a region. In such a plot, the x-axis typically represents the months of the year, while the y-axis shows the amount of rainfall received in millimeters. The plot may include multiple lines representing different years of data or may show a single line representing the long-term average rainfall for the region. By examining annual rainfall plots, one can identify patterns and trends in the rainfall data for a region. For example, a region may experience a consistently high amount of rainfall during a certain month or season, or may exhibit long-term trends of increasing or decreasing rainfall over time.

Annual rainfall plots (Fig. 6) are commonly used by climatologists, meteorologists, and hydrologists to analyze rainfall patterns and trends, as well as by farmers, water resource managers, and other stakeholders to make decisions related to agriculture, water management, and disaster planning.

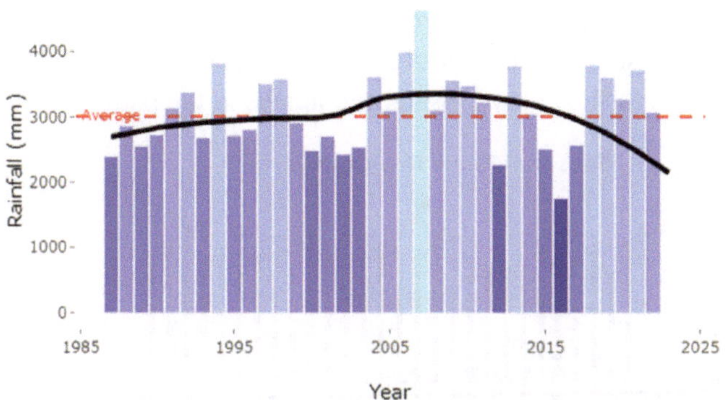

Fig. 6: Annual rainfall.

R Code for annual rainfall construction
#Load libraries after installation of packages

```
library(ggplot2)
library(plotly)
plot <- ggplot(data, aes(x = Year, y = Total_Rainfall, fill = Total_Rainfall)) + geom_col() +
scale_fill_gradient(low = "blue", high = "lightblue") +
geom_hline(yintercept = avg_rainfall, color = "red", linetype = "dashed") +
annotate("text", x = min_year, y = avg_rainfall, label = "Average", color = "red", size = 3) +
labs(title = "Annual Rainfall", x = "Year", y = "Rainfall (mm)", size = 14) +
theme(plot.background = element_rect(fill = "white", color = NA), panel.background =
element_blank(), plot.title = element_text(size = 20), legend.position = "none", plot.margin =
unit(c(1,1,1,1), "cm")) + stat_smooth(method = "loess", formula = y ~ x, se = FALSE, color
= "black")
ggplotly(plot)
```

Seasonal Rainy Day Count Plot

A seasonal rainy day count plot (Fig. 7) shows the number of rainy days in different seasons of the year. This type of plot can help to identify patterns in the occurrence of rainfall throughout the year.

To create a seasonal rainy day count plot, you first need to get the daily rainfall data for a particular location. Then, you can use the data to calculate the number of rainy days in each season. Seasons can be defined in different ways depending on the location, but commonly used definitions include meteorological seasons (winter: December–February, Summer: March–May, Southwest monsoon: June–August, Northeast monsoon: September–November). Once you have calculated the number of rainy days in each season, you can create a bar chart or a line chart to visualize the data. The x-axis would represent the different seasons and the y-axis would represent the number of rainy days. You can also add a trend line or a moving average to show any long-term changes in the pattern of rainfall.

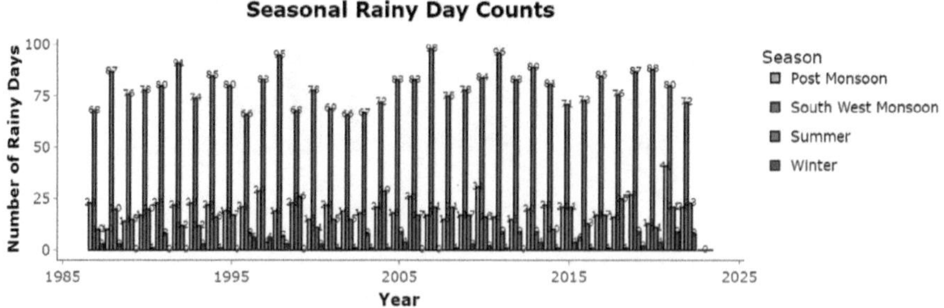

Fig. 7: A seasonal rainy day count plot.

```
# Seasonal rainy day plot
#Load libraries after installation of packages
library(ggplot2)
library(plotly)
# Aggregate the data by season and year
#The second section of the code aggregates the data by season and year. It calculates the number of rainy days for each season and year, and adds a new column to the data frame for the rainy day counts.
agg_data <- filtered_data() %>%
mutate(Year = year(Date),
 Rainy_Day = ifelse(rainfall > 2.5, 1, 0)) %>%
group_by(Season, Year) %>%
summarise(Rainy_Days = sum(Rainy_Day),
Total_Days = n()) %>%
ungroup() %>%
mutate(Rainy_Day_Count = Rainy_Days) # add a new column for rainy day counts
```

```
# Create the ggplot
p <- ggplot(agg_data, aes(x = Year, y = Rainy_Day_Count, fill = Season)) +
  geom_bar(stat = "identity", position = "dodge", color = "black") +
  geom_text(aes(label = Rainy_Day_Count), position = position_dodge(width = 0.9),
            vjust = -0.5, size = 3.5) + # add text labels on top of bars
  labs(title = "Seasonal Rainy Day Counts", x = "Year", y = "Number of Rainy Days",
       fill = "Season") +
  theme_classic(base_size = 14) +
  theme(panel.grid.major = element_blank(), panel.grid.minor = element_blank(),
        legend.position = "bottom", legend.direction = "horizontal",
        plot.title = element_text(hjust = 0.5, size = 20, face = "bold"),
        axis.title = element_text(size = 16, face = "bold"),
        axis.text = element_text(size = 14),
        legend.text = element_text(size = 14)) +
  scale_fill_brewer(palette = "Dark2") + # change color palette
  theme(plot.background = element_rect(fill = "white"), # add white background
        panel.background = element_rect(fill = "white"), # add white panel background
        legend.box.background = element_rect(color = "black"), # add border around legend
        legend.title = element_blank(), # remove legend title
        legend.margin = margin(t = 5, r = 5, b = 5, l = 5), # adjust legend margin
        legend.text.align = 0, # left-align legend text
        legend.spacing.x = unit(0.2, "cm")) # adjust legend spacing
# Convert ggplot object to plotly object
ggplotly(p)
```

The Rainfall Seasonal Decomposition Plot

The rainfall seasonal decomposition plot (Fig. 8) shows the various components of a time series of rainfall data. It is a useful tool for understanding the patterns and trends in the data. The plot typically displays the original data, the seasonal component, the trend component, and the random component.

The original data represents the observed rainfall values at each time point. The seasonal component denotes the periodic fluctuations in the data, such as the yearly or monthly cycles in rainfall. The trend component represents the long-term changes in the data, such as a gradual increase or decrease in rainfall over time. The random component represents the unexplained variation in the data which is not accounted for by the seasonal or trend components. The plot typically uses line graphs to display each of these components, with the x-axis representing time and the y-axis rainfall values. The different components are usually displayed in different colors to make it easy to distinguish between them.

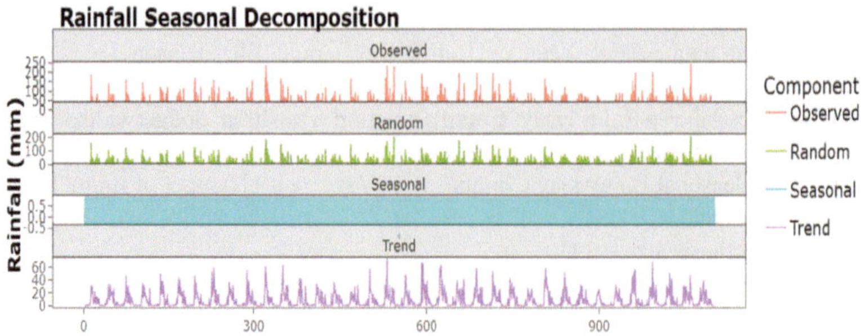

Fig. 8: Rainfall seasonal decomposition plot.

```
#Load libraries after installation of packages
library(ggplot2)
library(plotly)
# Create time series data
ts_data <- ts(filtered_data()$rainfall, frequency = 12)
# Perform seasonal decomposition
decomp <- decompose(ts_data)
# Convert decomposed data into a data frame
df <- data.frame(
 Date = time(ts_data),
 Observed = decomp$x,
 Seasonal = decomp$seasonal,
 Trend = decomp$trend,
 Random = decomp$random)
# Reshape data frame to long format
df_long <- tidyr::pivot_longer(df, cols = -Date, names_to = "Component", values_to =
"Value")
# Create plot using ggplot
p <- ggplot(df_long, aes(x = Date, y = Value, color = Component)) +
 geom_line() +
 facet_wrap(~ Component, ncol = 1, scales = "free_y") +
 labs(title = "Rainfall Seasonal Decomposition", x = "", y = "Rainfall (mm)",
     color = "Component") +
 theme_bw(base_size = 16) +
 theme(panel.grid.major = element_blank(), panel.grid.minor = element_blank(),
     legend.title = element_text(size = 18),
     legend.text = element_text(size = 16),
     plot.title = element_text(size = 20, face = "bold"),
     axis.title = element_text(size = 18, face = "bold"),
     axis.text = element_text(size = 12))
# Convert ggplot object to plotly object
ggplotly(p)
```

Boxplot for Seasonal Rainfall Analysis

A boxplot (Fig. 9) is a graph that displays the distribution of a dataset through its quartiles. It shows the median value, upper and lower quartiles, and any outliers that may be present. In the given code, a boxplot is used to denote the seasonal rainfall data. Each box represents the interquartile range (IQR) of the rainfall data for a particular season. The IQR is the range between the first quartile (25th percentile) and the third quartile (75th percentile) of the data. The vertical line inside the box represents the median rainfall value.

Furthermore, the code includes individual data points as small circles on the boxplot. This provides a more detailed view of the distribution of data points within each season. By filling the boxes with different colors corresponding to the different seasons, the plot allows easy comparison of rainfall distribution across seasons. The legend at the bottom of the plot provides clarity on which color represents which season. Overall, the boxplot is an effective way to visualize and compare the distribution of rainfall across seasons in a clear and concise manner.

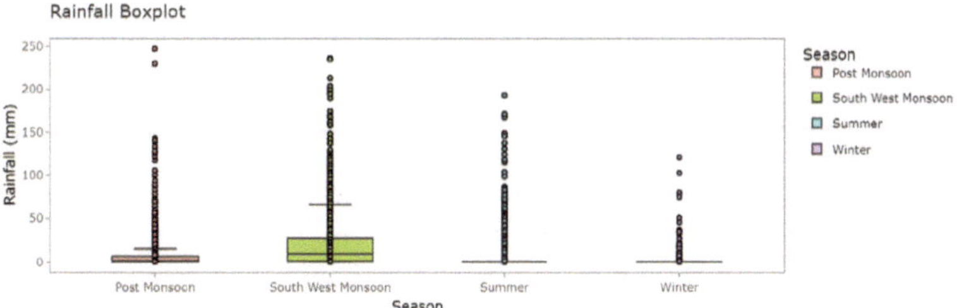

Fig. 9: Box plot.

```
#Load libraries for boxplot preparation
library(ggplot2)
library(plotly)
#Boxplot code
p <- ggplot(filtered_data(), aes(Season, rainfall, fill = Season)) +
   geom_boxplot(alpha = 0.7, outlier.color = NA) +
   geom_point() +
   labs(title = "Rainfall Boxplot", x = "Season", y = "Rainfall (mm)", fill = "Season") +
   theme_minimal() +
   theme(legend.position = "bottom")
ggplotly(p)
```

Conclusion

In conclusion, hydrological data visualization is a crucial tool for analyzing and understanding water resources. R Language provides several powerful packages such as ggplot2, plotly, and hydroTSM that can be used for creating static and dynamic plots to visualize different hydrological variables. Each package has its own strengths and limitations, and choosing the appropriate package and plot type is essential for gaining insight into the patterns and trends in the data. By leveraging these tools, we can use them efficiently in decision-making of water resources management, flood forecasting, and climate adaptation, ultimately contributing to the conservation and sustainability of water resources.

References

Sievert, C. Interactive data visualization with R, plotly, and shiny. CRC Press, 2020.

Wickham, H. ggplot2: Elegant Graphics for Data Analysis. Springer, 2016.

Zambrano-Bigiarini, M. Tutorial for Introductory Analysis of Daily Precipitation Data with hydroTSM, 2017.

Section 2

Development of Climate and Drought Indices

4

Computation of Potential, Actual, and Reference Evapotranspiration Using R Package *Evapotranspiration*

***Fathima Sona N*[1]*, Naveena K*[1]*, and Surendran U*[1]**

[1]*Centre for Water Resources and Development, Kunnamangalam, Kozhikode*

Abstract

Evapotranspiration is a crucial process in the Earth's water cycle, which involves water loss from soil, plants, and open water surfaces through evaporation and transpiration. It plays a vital role in plant growth and water resource management making accurate estimation of evapotranspiration essential for crop management, irrigation planning, and water resource management. However, direct measurement of evapotranspiration can be a challenging and time-consuming process, which is why identifying the most effective empirical evapotranspiration method is necessary. There are several methods available for estimating evapotranspiration, including lysimeter data, water budgeting, empirical models, pan evaporimeters, and combination theory. Empirical models based on radiation, temperature, mass transfer, and combination theory are also widely used for estimating evapotranspiration. One of the most efficient methods for calculating evapotranspiration is the use of the R software package, Evapotranspiration, which offers 21 models for estimating potential, actual, and reference evapotranspiration. The package can produce both average and cumulative values, providing valuable information for decision-making. In conclusion, the accurate estimation of evapotranspiration is crucial for effective crop management, irrigation planning, and water resource management. The use of empirical models, such as those available in the Evapotranspiration package, can provide efficient and effective estimates of evapotranspiration, enabling researchers and practitioners to better understand and manage evapotranspiration processes.

Keywords: *Empirical models, evaporation, evapotranspiration, R package, water cycle*

Introduction

Evapotranspiration is a vital process in the Earth's water cycle, and it involves two main processes—evaporation and transpiration. Evaporation refers to the conversion of water from its liquid state to its gaseous state, which occurs primarily from the surface of bodies of water, wet vegetation, and soils. This process requires energy, which is typically provided by direct solar radiation and ambient air temperature. Transpiration, on the other hand, is the process by which plants release water vapor through small openings in their leaves called stomata. This process is essential for plant growth and function as it helps regulate the plant's temperature and allows for nutrient uptake and water transport. Like evaporation, transpiration also requires energy, with factors like vapor pressure gradient and wind speed influencing the rate at which water is released.

To accurately measure evapotranspiration, it is necessary to consider a range of factors such as temperature, radiation, wind speed, and humidity. Plant parameters, such as the presence of a cuticle or the size and number of stomata, can also greatly influence the transpiration rate. Other factors that may impact evapotranspiration include the season, time of day, and soil properties such as water holding capacity and root depth. Evapotranspiration is a complex concept that has been studied by various scientists and organizations over time. Three main evapotranspiration (ET) concepts have been developed: Reference ET, ET under standard conditions, and ET under nonstandard conditions. The goal of determining potential or reference evapotranspiration is to exclude crop-specific variations from the evapotranspiration process. The potential evapotranspiration concept, introduced in the late 1940s and 50s by Penman, assumes constant crop conditions and defines it as "the amount of water transpired in a given time by a short green crop, completely shading the ground, of uniform height and with adequate water status in the soil profile." ETo, which is a climatic parameter, represents the evaporating power of the atmosphere at a certain location and time of the year and is computed from weather data. It does not take into account crop characteristics or soil conditions, and the only factors affecting it are climatic parameters. The range of ETo varies from 1–2 mm/day in humid and subhumid temperate regions to 6–9 mm/day in warm arid and semi-arid temperate regions and from 2–3 in cool humid and subhumid tropics to 6–8 mm/day in warm arid and semi-arid tropics. Actual ET is the amount of water on a large area of grass that is actively growing and covering the entire soil surface it actually uses. The amount of water that is actually evacuated from a surface as a result of the processes E and T is known as ETa. Crop water needs can be calculated by the following equation.

Crop water needs = potential evapotranspiration - actual evapotranspiration

Various methods are used for estimating evapotranspiration, including lysimeter data, water budgeting, empirical models, pan evaporimeters, and other techniques (Ghiat *et al*., 2021). Lysimeter data is considered the most accurate, but it can be expensive and labor-intensive. Other methods rely on empirical formulations based on factors such as radiation, temperature, and mass transfer. Different equations have been developed by researchers worldwide, but none of them is perfect. To address this, the American Society of Civil Engineers (ASCE) and the Food and Agriculture Organization (FAO) developed the Penman-monteith equation in 1956 for calculating Reference Evapotranspiration (Poyen *et al*., 2016).

Methodology

R Package Evapotranspiration

The R Package Evapotranspiration is a powerful tool that includes 21 different formulations, such as Penman, Penman-monteith FAO-56, Priestley Taylor, and Morton formulations. These models are based on two fundamental components that drive ET, namely energy balance and mass transfer. The energy balance determines the latent heat of vaporization, which can be estimated using the total incoming solar radiation (Rs), net incoming radiation (R), sensible heat exchange (H), heat exchange with the ground (G), and net input of water advected energy (Ad) (Guo *et al*., 2016). Meanwhile, mass transfer influences the rate of movement of water vapor away from the evaporating surface, and is impacted by the vapor gradient (related to relative humidity [RH] and temperature [T]) and wind speed uz. The aerodynamic and surface resistances also play critical roles in regulating the evaporation flux, with the latter controlling stomatal opening in plant leaves (Monteith, 1991). When combined, these components provide a comprehensive understanding of ET and its associated variables, such as temperature, relative humidity, incoming solar radiation, and wind speed. The models available in the R Package Evapotranspiration are listed in Table 1, making it an indispensable tool for researchers and practitioners seeking to better understand and manage ET processes.

Table 1: Models considered in Evapotranspiration package

Model	**Equation**
Abtew	Actual evapotranspiration
Blaney Cridle	Reference crop evapotranspiration
BrutsaertStrickler	Actual evapotranspiration
ChapmanAustralian	Potential evapotranspiration

GrangerGray	Actual evapotranspiration
Hamon	Potential evapotranspiration
HargreavesSamani	Reference crop evapotranspiration
JensenHaise	Potential evapotranspiration
Linacre	Actual evapotranspiration
Makkink	Reference crop evapotranspiration
MattShuttleworth	Reference crop evapotranspiration
McGuinnessBordne	Potential evapotranspiration
MortonCRAE	Potential evapotranspiration, wet-environment areal evapotranspiration, and actual areal evapotranspiration
MortonCRWE	Potential evapotranspiration or shallow lake evaporation.
Penman	Potential evapotranspiration
PenmanMonteith	Reference crop evapotranspiration
PenPan	Class-A pan evaporation
PriestleyTaylor	Potential evapotranspiration
Romanenko	Potential evapotranspiration
SzilagyiJozsa	Actual evapotranspiration
Turc	Reference crop evapotranspiration

Preparing Input Data

Before estimating evapotranspiration using the R Package Evapotranspiration, input data must be prepared. The input data required for the calculation of evapotranspiration includes the year, month, and day of the record, as well as various climate parameters such as maximum temperature, minimum temperature, relative humidity, wind speed, and sunshine hours. The specific input data required may vary depending on the model used, but a comprehensive data set containing all the basic climate parameters is preferred.

In addition to input data, the R Package Evapotranspiration requires constants for the calculation of evapotranspiration. The "constants" data set includes both universal constants that remain unchanged under most conditions and location-specific constants. This data set includes 36 constants, with 20 universal constants and 16 constants specific to the Kozhikode region of Kerala, India, as indicated in Table 2. By providing accurate input data and constants, the R Package Evapotranspiration can effectively estimate evapotranspiration and aid in crop management, irrigation planning, and water resource management.

Table 2: Universal constants for different models in the Evapotranspiration package

	Constant	Value
lambda	Latent heat of evaporation	2.45 $MJ.kg^{-1}$ at 20°C
sigma	Stefan-Boltzmann constant	4.903 ′ 10^{-9} $MJ.K^{-4}.m^{-2}.day^{-1}$
Gsc	Solar constant	0.0820 $MJ.m^{-2}min^{-1}$
Roua	Mean density of air	1.2 $kg.m^{-3}$ at 20°C
Ca	Specific heat of air	0.001013 $MJ.kg^{-1}.K^{-1}$
G	Soil heat flux negligible for daily time-step	0 (Allen *et al.*, 1998)
alphaA	Albedo for Class-A pan	0.14
alphaPT	Priestley-Taylor coefficient	1.26 for Priestley-Taylor formula (Priestley and Taylor, 1972) 1.31 for Szilagyi-Jozsa formula (Szilagyi and Jozsa, 2008) 1.28 for Brutsaert-Strickler formula (Brutsaert and Strickler, 1979)
ap	Constant in Penpan	2.4
b0	Constant in Morton's procedure	1
b1	Constant in Morton's procedure	14 $W.m^{-2}$ (Chiew and McMahon, 1991)
b2	Constant in Morton's procedure	1.2 (Chiew and McMahon, 1991)
e0	Constant for Blaney-Criddle formula	0.81917 (Frevert *et al.*, 1983)
e1	Constant for Blaney-Criddle formula	-0.0040922 (Frevert *et al.*, 1983)
e2	Constant for Blaney-Criddle formula	1.0705 (Frevert *et al.*, 1983)
e3	Constant for Blaney-Criddle formula	0.065649 (Frevert *et al.*, 1983)
e4	Constant for Blaney-Criddle formula	= -0.0059864 (Frevert *et al.*, 1983)
e5	Constant for Blaney-Criddle formula	-0.0005967 (Frevert *et al.*, 1983)
EpsilonMo	Land surface emissivity in Morton's procedure	0.92,
sigmaMo	Stefan-Boltzmann constant in Morton's procedure	5.67e-08 $W.m^{-2}.K^{-4}$.

Table 3: Sixteen Variable constants specific for the Kozhikode region

	Constant	Value
Latitude	Latitude radians for study location,	11.24
lat_rad	Latitude in radians	0.196
as	Fraction of extraterrestrial radiation reaching Earth on sunless days	0.25

	Difference between the fraction of extraterrestrial radiation reaching full-sun days and that on sunless days	0.5
Elev	Ground elevation above mean sea level	60 m
z	Height of wind instrument	3 m
fz	Constant in Morton's procedure	= 28.0 $W.m^{-2}.mbar^{-1}$ for CRAE model for T > 0°C = 25.0 $W.m^{-2}.mbar^{-1}$ for CRWE model for T > 0°C (Morton, 1983)
a_0	Constant for estimating sunshine hours from cloud cover data	
b_0	Constant for estimating sunshine hours from cloud cover data	
c_0	Constant for estimating sunshine hours from cloud cover data	
d_0	Constant for estimating sunshine hours from cloud cover data	
gammaps	Product of Psychrometric constant and atmospheric pressure as sea level (mbar. $°C^{-1}$)	0.67
PA	Annual precipitation (mm) Use the actual annual precipitation of case study	3,353 mm
alphaMo	Constant in Morton's procedure (dimensionless)	17.27
betaMo	Constant in Morton's procedure (°C). Two alternative values depending on the baseline temperature at study site	237.3 when T > 0°C 237.3 for Adelaide 40 265.5 when T < 0°C (Morton, 1983a)
lambdaMo	Latent heat of vaporization in Morton's procedure ($W.day.kg^{-1}$). Two alternative values depending on the baseline temperature at study site:	28.5 when T > 0°C 28.5 × 1.15 when T < 0°C (Morton, 1983a)

R Codes for ET Calculation

```
#Load the packages into the current R session
library(Evapotranspiration) library(zoo)
#Import the input data from a CSV file
calicut_data <- read.csv(file.choose(), header = TRUE)
#Edit the constants based on the location
constants_data <- edit(constants)
#Process the input data using the ReadInputs function
processed_data <- ReadInputs(varnames = c("Year", "Month", "Day", "Tmax", "Tmin",
"RHmax", "RHmin", "uz", "n"), data = calicut_data, con = constants_data, stopmissing =
c(10, 10, 3), timestep = "daily", interp_missing_days = FALSE, interp_missing_entries =
FALSE, interp_abnormal = FALSE, missing_method = NULL, abnormal_method = NULL)
#Calculate ET using the ET.Abtew function
ET_result <- ET.Abtew(processed_data, con = constants_data, ts = "daily", solar = "sunshine
hours", message = "yes", AdditionalStats = "yes", save.csv = "yes")
#View the ET results
ET_result
```

The function ET.Abtew is used to calculate evapotranspiration using the Abtew method. The required input parameters are the climate variables Tmax, Tmin, Rs or n or Cd (depending on the availability of the data) stored in a list named Cal, and a list of constants required for the calculation stored in a list named Con. The desired time step of the output ET estimates can be specified using the argument ts, which can be set as daily, monthly, or annual. The argument solar can be used to specify how solar radiation is calculated and can be set as data, sunshine hours, cloud, or monthly precipitation. The default value for solar is sunshine hours.

message: Must be either yes or no, indicating whether message should be printed for calculation summary including the following elements: - ET model name and ET quantity estimated - Option for calculating solar radiation (i.e., the value of argument solar) - Time step of the output ET estimates (i.e., the value of argument ts) - Units of the output ET estimates - Time duration of the ET estimation - Number of ET estimates obtained in the entire time-series - Basic statistics of the estimated ET time-series including mean, max and min values.

AdditionalStats: "yes" or "no" indicating whether monthly averaged and annual averaged ET should be calculated.

save.csv: Must be either yes or no, indicating whether a .csv of ET estimates should be saved to working directory: Dummy for generic function, no need to define

The other methods also can be worked out similarly with due changes in constants and other commands.

```
ET.Abtew(Cal,con,ts="daily",solar = "sunshine hours", message = "yes", AdditionalStats = "yes",save.csv = "yes")
ET.BlaneyCriddle(Cal,con,ts="daily",solar = "sunshine hours", height= F,message = "yes", AdditionalStats = "yes",save.csv = "yes")
ET.HargreavesSamani(Cal,con,ts="daily",message = "yes", AdditionalStats = "yes",save.csv = "yes")
ET.JensenHaise(Cal,con,ts="daily",solar = "sunshine hours", message = "yes", AdditionalStats = "yes",save.csv = "yes")
ET.Makkink(data1,con,ts="daily",solar="sunshine hours", message="yes",AdditionalStats = "yes",save.csv = "yes")
ET.Penman(data1,con,ts="daily",solar = "sunshine hours", windfunction_ver = "1948", alpha = 0.08, z0 = 0.001,message="yes", AdditionalStats="yes", save.csv="yes")
ET.PenmanMonteith(data1,con,ts="daily",solar = "sunshine hours", wind="yes", crop= "short", message="yes", AdditionalStats="yes", save.csv="yes")
ET.PriestleyTaylor(data1,con,ts="daily",solar = "sunshine hours", alpha= 0.23, message="yes", AdditionalStats="yes", save.csv="yes")
ET.BrutsaertStrickler(data1,con,ts="daily",solar = "sunshine hours",alpha = 0.25,message = "yes",AdditionalStats ="yes",save.csv = "yes" )
ET.ChapmanAustralian(data1,con,ts="daily",PenPan=T,solar = "sunshine hours",alpha = 0.25,message = "yes",AdditionalStats ="yes",save.csv = "yes" )
ET.GrangerGray(data1,con,ts="daily",PenPan=T,solar = "sunshine hours",windfunction_ver=1948, alpha = 0.25,message = "yes",AdditionalStats ="yes",save.csv = "yes" )
```

Here, **Cal** represents the input data in a list format with required climate variables, and **con** represents the constants required for each method. The methods can be chosen based on the ET formula to be used, and the desired time step (**ts**) and solar radiation type (**solar**) can be specified as well. Additionally, various other options like additional statistics and saving output data as CSV can also be specified.

Illustrations

A case has been used to demonstrate the core utilities of the package Evapotranspiration, using climate data from meteorological sites at Kottamparamba (11.17° N, 75.52° E) in Kozhikode, Kerala, for the period from 01/01/2019 to 12/31/2019. The output is obtained as monthly average and cumulative evapotranspiration values, which are represented below (Tables 4 and 5).

Table 4: Monthly cumulative evapotranspiration

ET.Monthly	PM-FAO	Penman	Penpan	Turc	Romanenko	MattShuttleworth	GrangerGray	Abtew
Jan-19	90.4	135.4	132.3	98.5	138.4	68.4	69.5	129.1
Feb-19	91.4	135.3	132.7	98.0	131.9	70.6	72.0	125.5
Mar-19	116.5	169.7	170.3	120.8	157.0	88.1	89.6	147.5
Apr-19	109.0	157.0	157.1	116.2	128.7	89.4	91.8	136.1
May-19	103.4	148.2	147.9	111.5	110.5	87.4	91.0	129.1
Jun-19	80.0	118.9	116.8	90.1	60.2	68.7	78.1	109.0
Jul-19	70.7	105.5	104.2	75.6	56.8	57.6	67.1	90.9
Aug-19	71.7	106.6	104.2	79.6	50.9	61.6	71.2	95.9
Sep-19	76.6	110.4	107.1	84.9	63.0	67.8	73.9	102.5
Oct-19	76.4	109.8	106.1	85.2	74.7	67.3	71.6	103.6
Nov-19	70.1	102.5	99.2	80.7	69.1	60.5	65.1	101.2
Dec-19	79.8	118.6	116.3	88.6	109.0	62.0	64.6	113.8

Table 4.1: Monthly cumulative evapotranspiration (Continued)

ET.Monthly	Hargreves Samani	Jensen Haise	Makkink	Priestley Taylor	GragerGrey	BrutsaertStrickler	SzilagyiJozsa
Jan-19	87.9	101.6	87.0	94.6	69.5	77.5	95.1
Feb-19	86.4	103.7	86.5	97.6	72.0	83.4	102.5
Mar-19	104.5	133.0	105.5	123.6	89.6	106.7	128.7
Apr-19	107.4	131.0	99.8	123.4	91.8	116.8	141.4
May-19	110.4	125.2	94.6	120.5	91.0	118.1	143.1

Jun-19	103.5	95.5	76.2	100.4	78.1	102.8	88.7
Jul-19	106.6	77.1	62.0	86.6	67.1	85.5	102.0
Aug-19	107.1	81.9	65.8	90.9	71.2	93.6	113.4
Sep-19	100.1	89.0	71.2	94.6	73.9	97.7	121.7
Oct-19	95.8	88.5	71.4	92.0	71.6	93.0	116.0
Nov-19	82.4	82.4	68.4	83.7	65.1	83.0	103.2
Dec-19	80.5	90.6	76.6	86.3	64.6	75.0	91.6

Table 5: Monthly average evapotranspiration

ET.Monthly	PM-FAO	Penman	Penpan	Turc	Romanenko	MattShuttleworth	GrangerGray	Abtew
Jan-19	2.92	4.37	4.27	3.18	4.46	2.21	2.24	4.16
Feb-19	3.26	4.83	4.74	3.50	4.71	2.52	2.57	4.48
Mar-19	3.76	5.47	5.49	3.90	5.06	2.84	2.89	4.76
Apr-19	3.63	5.23	5.24	3.87	4.29	2.98	3.06	4.54
May-19	3.34	4.78	4.77	3.60	3.56	2.82	2.93	4.16
Jun-19	2.67	3.96	3.89	3.00	2.01	2.29	2.60	3.63
Jul-19	2.28	3.40	3.36	2.44	1.83	1.86	2.16	2.93
Aug-19	2.31	3.44	3.36	2.57	1.64	1.99	2.30	3.09
Sep-19	2.55	3.68	3.57	2.83	2.10	2.26	2.46	3.42
Oct-19	2.46	3.54	3.42	2.75	2.41	2.17	2.31	3.34
Nov-19	2.34	3.42	3.31	2.69	2.30	2.02	2.17	3.37
Dec-19	2.58	3.83	3.75	2.86	3.52	2.00	2.08	3.67

Table 5: Monthly average evapotranspiration (Continued)

ET.Monthly	Hargreves Samani	Jensen Haise	Makkink	Priestley Taylor	GragerGrey	BrutsaertStrickler	SzilagyiJozsa
Jan-19	1.43	2.84	3.28	2.81	3.05	2.50	3.069
Feb-19	1.59	3.08	3.70	3.09	3.49	2.98	3.661
Mar-19	1.76	3.37	4.29	3.40	3.99	3.44	4.152
Apr-19	1.19	3.58	4.37	3.33	4.11	3.89	4.713
May-19	0.68	3.56	4.04	3.05	3.89	3.81	4.615
Jun-19	0.15	3.45	3.18	2.54	3.35	3.43	2.958
Jul-19	−0.62	3.44	2.49	2.00	2.79	2.76	3.290
Aug-19	−0.54	3.45	2.64	2.12	2.93	3.02	3.659
Sep-19	−0.28	3.34	2.97	2.37	3.15	3.26	4.055
Jan-19	1.43	2.84	3.28	2.81	3.05	2.50	3.069
Feb-19	1.59	3.08	3.70	3.09	3.49	2.98	3.661
Mar-19	1.76	3.37	4.29	3.40	3.99	3.44	4.152
Apr-19	1.19	3.58	4.37	3.33	4.11	3.89	4.713
May-19	0.68	3.56	4.04	3.05	3.89	3.81	4.615
Jun-19	0.15	3.45	3.18	2.54	3.35	3.43	2.958
Jul-19	-0.62	3.44	2.49	2.00	2.79	2.76	3.290
Aug-19	-0.54	3.45	2.64	2.12	2.93	3.02	3.659
Sep-19	-0.28	3.34	2.97	2.37	3.15	3.26	4.055
Oct-19	-0.20	3.09	2.86	2.30	2.97	3.00	3.742
Nov-19	0.04	2.75	2.75	2.28	2.79	2.77	3.439
Dec-19	0.70	2.60	2.92	2.47	2.78	2.42	2.956

Conclusion

To summarize, evapotranspiration is a crucial process in the Earth's water cycle, which has a significant influence on plant growth and water resource management. Direct measurement of evapotranspiration can be challenging and time-consuming, making the use of empirical models essential. The R software package, Evapotranspiration, offers 21 models for estimating actual, potential, and reference evapotranspiration, which can produce both cumulative and average values. These estimates are valuable for decision-making in crop management, irrigation planning, and water resource management. The use of empirical models, such as those available in the Evapotranspiration package, provides efficient and effective estimates of evapotranspiration, enabling researchers and practitioners to better understand and proper management of evapotranspiration processes.

References

Allen, R.G., Pereira, L.S., Raes, D., Smith, M. "Crop evapotranspiration-Guidelines for computing crop water requirements-FAO Irrigation and drainage paper 56." FAO, Rome 300 no. 9 (1998): D05109.

Brutsaert, W., Stricker, H. "Asn Advection-Aridity Approach to Estimate Actual Regional Evapotranspiration." Water Resources Research 15, no. 2 (1979): 443–50.

Chiew, F.H., McMahon, T.A. "The Applicability of Morton's And Penman's Evapotranspiration Estimates in Rainfall-Runoff Modeling 1." JAWRA Journal of the American Water Resources Association 27, no. 4 (1991): 611–20.

Frevert, D.K., Hill, R.W., Braaten, B. C. "Estimation of FAO Evapotranspiration Coefficients." Journal of Irrigation and Drainage Engineering 109, no. 2 (1983): 265-70.

Ghiat, I., Mackey, H.R., Al-Ansari, T. "A Review of Evapotranspiration Measurement Models, Techniques and Methods for Open and Closed Agricultural Field Applications." Water 13, no. 18 (2021): 2523.

Guo, D., Westra, S., Maier, H.R. "An R Package for Modelling Actual, Potential and Reference Evapotranspiration." Environmental Modelling & Software, 78 (2016): 216–24.

Irmak, S., Haman, D. Z. Evapotranspiration: Potential or Reference? ABE 343/AE256, 6/2003. EDIS, 2003(14).

Poyen, E.F.B., Ghosh, A.K., PalashKundu, P. "Review on Different Evapotranspiration Empirical Equations." International Journal of Advanced Engineering, Management and Science 2, no. 3 (2016); 239382.

Szilagyi, J., Jozsa, J. (2008). "New findings About the Complementary Relationship-Based Evaporation Estimation Methods." Journal of Hydrology 354, no. 1–4 (2008): 171–86.

5

Comparison of Meteorological Drought Indices for Assessment of Drought Features

Santhosh Onte, Naveena K[1], Devika Sajan[1], Surendran U[1] and K Ch V Naga Kumar

[1]*Centre for Water Resources Development and Management, Kunnamangalam Kozhikode*

Abstract

Drought is a natural phenomenon that occurs due to a significant lack of precipitation and can cause extensive damage and impact a significant number of people. To minimize the damage caused by drought, it is important to characterize and quantify the severity of drought. Several drought indices have been developed for this purpose, with many relying solely on precipitation data or using a combination of meteorological variables. This study focuses on three commonly used drought indices, namely the Rainfall Anomaly Index (RAI), Standardized Precipitation Index (SPI), and Standardized Precipitation Evapotranspiration Index (SPEI). SPI and RAI are calculated using only rainfall data whereas SPEI utilizes data on rainfall, maximum temperature, and minimum temperature. The study employs weather data from the humid tropical region of Kerala State, India, where rainfall is poorly distributed, with an average annual rainfall of above 3,000 mm. The results reveal that the drought classification based on different indices is not unique, with SPEI (208) and RAI (262) exhibiting more metrological drought days compared to SPI (49). As SPEI incorporates temperature data into its index development, it is deemed more appropriate for characterizing drought severity in the study area compared to the other indices considered.

Keywords: Drought; Rainfall anomaly index (RAI), standardized precipitation index (SPI), standardized precipitation evapotranspiration index (SPEI)

Introduction

Droughts are considered to be a significant environmental disaster and have captured the attention of various experts, including environmentalists, ecologists, hydrologists, meteorologists, geologists, and agricultural scientists. These natural calamities are known to occur in almost all climatic zones, regardless of whether they are high- or low-rainfall areas. Generally a decrease in the amount of precipitation received over an extended period such as a season or a year, is the primary cause of droughts. However, other factors such as high temperatures, strong winds, low relative humidity, and characteristics of rainfall (e.g., intensity, duration, onset, and termination) also play a crucial role in their occurrence. It's important to distinguish a drought from aridity, which is a permanent feature of a climate and is limited to low rainfall areas (Wilhite, 1992). A drought is a temporary phenomenon and can last from a few months to several years. Some people often mistake droughts for heat waves, but Chang and Wallace (1987) highlighted the fundamental difference between these two events. A heat wave typically lasts for a week, while a drought can extend for several months or even years. When a heat wave and a drought coincide, it can lead to dire socio-economic consequences.

The water demand has increased manyfold as a result of population growth and the expansion of agricultural, energy, and industrial sectors, and water scarcity occurs almost every year in many parts of the world. Other factors, such as climate change and water contamination, have also contributed to water scarcity. Floods and droughts have had higher peaks and severity levels in recent years. In some areas, the time between extreme events appears to be getting shorter. This shift in the occurrence of extreme hydrologic events has been mentioned by Lettenmaier *et al.*, (1996) and Aswathanarayana (2001).

Droughts have an impact on both surface and groundwater resources, resulting in reduced water supply, degraded water quality, crop failure, reduced range productivity, reduced power generation, disturbed riparian habitats, and suspended recreation activities, among other things (Riebsame *et al.*, 1991). Droughts also have an impact on water quality because moderate climate fluctuations alter hydrologic regimes, which have a significant impact on lake chemistry (Webster *et al.*, 1996). Runoff transports sediment, organic matter, and nutrients to surface waters, a pathway that is disrupted during droughts.

What is Drought?

It is difficult to provide a precise and universally accepted definition for drought due to its varying characteristics and impacts across different regions of the world, such as rainfall patterns, human response and resilience, and diverse academic perspectives.

Drought is a deficiency in precipitation over an extended period, usually a season or more, resulting in a water shortage causing adverse impacts on vegetation, animals, and/or people. It is a normal, recurrent feature of climate that occurs in virtually all climate zones, from very wet to very dry. Drought is a temporary aberration from normal climatic conditions; thus it can vary significantly from one region to another. Drought is different than aridity, which is a permanent feature of climate in regions where low precipitation is the norm, as in a desert.

Some of the commonly used definitions (conceptual) are:

- The World Meteorological Organization (WMO, 1986)—"drought means a sustained, extended deficiency in precipitation."
- The Food and Agriculture Organization (FAO, 1983) of the United Nations—"the percentage of years when crops fail from the lack of moisture."
- The encyclopedia of climate and weather (Schneider, 1996)—"an extended period – a season, a year, or several years – of deficient rainfall relative to the statistical multi-year mean for a region."
- Palmer (1965)—"drought as a significant deviation from the normal hydrologic conditions of an area."
- Linseley *et al.*, (1959)—"drought as a sustained period of time without significant rainfall."

Classification of Drought

Figure 1 shows the different classifications of drought.

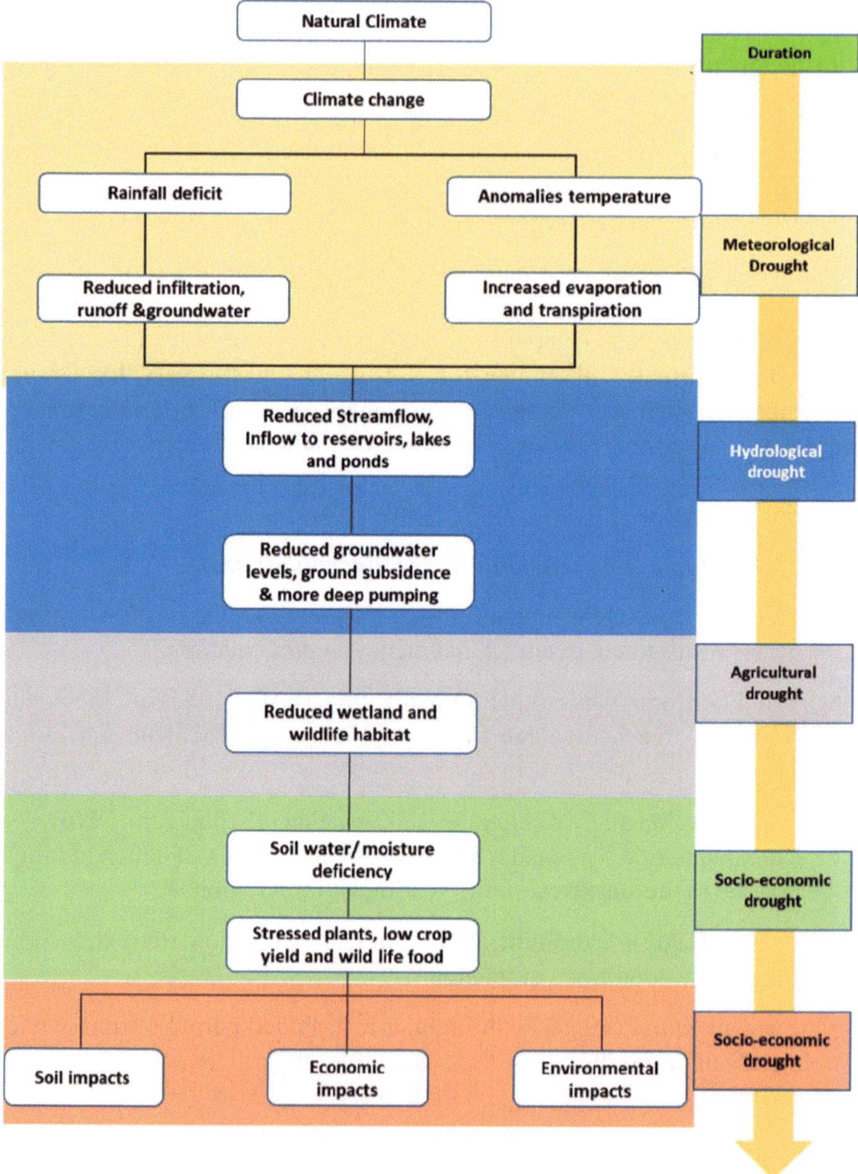

Fig. 1: Different classification of drought.

Meteorological Drought

Meteorological drought is usually defined based on the degree of dryness (in comparison to some "normal" or average) and the duration of the dry period. Drought onset generally occurs with a meteorological drought.

Hydrological Drought

Hydrological drought usually occurs following periods of extended precipitation shortfalls that impact water supply (i.e., streamflow, reservoir and lake levels, groundwater), potentially resulting in significant societal impacts. As regions are interconnected by hydrologic systems, the impact of meteorological drought may extend well beyond the borders of the precipitation-deficient area.

Agricultural Drought

Agricultural drought links various characteristics of meteorological (or hydrological) drought to agricultural impacts, focusing on precipitation shortages, soil water deficits, reduced groundwater or reservoir levels needed for irrigation, and so forth.

Socio-economic Drought

Socio-economic drought is associated with the failure of water resources systems to meet water demands and thus associating droughts with the supply of and demand for an economic good (water) (AMS, 2004). Socio-economic drought occurs when the demand for an economic good exceeds the supply as a result of a weather-related shortfall in water supply.

Several studies have discussed these four types of droughts; however, it will be useful and important to introduce groundwater drought as a type of drought that has not been included in the classification of droughts. To date, little research has been done on the occurrence and propagation of droughts in groundwater. The following section discusses groundwater drought in more detail as this can be treated as a new type of drought.

Groundwater Drought

When groundwater systems are affected by droughts, first groundwater recharge and later groundwater levels and groundwater discharge decrease. Such droughts are called groundwater droughts and generally occur on a time scale of months to years (van Lanen and Peters, 2000).

Drought Indices

Drought indices are mathematical tools used to measure the severity of drought, which are calculated and analyzed based on climatic or hydrometeorological inputs known as drought indicators. These indicators refer to various parameters, such as rainfall, temperature, streamflow, groundwater and reservoir levels, soil moisture, and snowpack, which are used to describe drought conditions (Handbook of Drought Indicators and Indices, 2016). To accurately measure drought, a range of drought indices have been developed. While some drought indices rely solely on precipitation, others use a combination of meteorological variables. Moreover, there have been numerous studies conducted to investigate the impact of climate change on the drying of terrestrial surfaces across the globe.

Methodology

The study at hand focuses on various drought indices, including the Standardized Precipitation Index (SPI), Standardized Precipitation Evapotranspiration Index (SPEI), and Rainfall Anomaly Index (RAI). These indices are particularly useful in assessing drought conditions and can be calculated using different meteorological variables. For example, SPI and RAI are calculated based solely on rainfall data, while SPEI requires data on rainfall, maximum temperature, and minimum temperature.

Standardized Precipitation Index (SPI)

The SPI is calculated using long-term rainfall or precipitation data, which are transformed into a normal distribution. The deviation of an observed value from the long-term mean is then measured in terms of the number of standard deviations for a normally distributed random variable (McKee *et al.*, 1993; Edwards and McKee 1997). A detailed method for determining SPI values is described by Tigkas *et al.*, 2014. The SPI classification, as shown in Table 1, is used to interpret the SPI values.

Table 1: Wet and drought period classification according to the SPI index.

SPI Range	Class
2.0 and above	Extremely wet
1.5 to 1.99	Very wet
1.0 to 1.49	Moderately wet
−0.99 to 0.99	Near normal
−1.0 to −1.49	Moderately dry
−1.5 to −1.99	Severely dry
−2 and less	Extremely dry

$$SPI_{ijk} = \frac{P_{ijk} - \overline{P_{ij}}}{\sigma_{ij}} \tag{1}$$

where SPI_{ijk} is the z-value for the station i during timeframe j for year k, P_{ijk} is the precipitation value for station i during timeframe j for year k, $\overline{P_{ij}}$ is the mean for station i during timeframe j over n years, and σ_{ijk} is the standard deviation of pixel i during week j over n years. SPI was calculated for different time scales 3, 6, and 12 months by fitting into the gamma distribution function.

Standardized Precipitation Evapotranspiration Index (SPEI)

The SPEI is an enhanced version of the SPI that incorporates both rainfall and potential evapotranspiration (PET) in identifying drought conditions. This makes the SPEI a more comprehensive tool for assessing the impact of rising temperatures on water availability. The SPEI can be calculated over various timescales ranging from 1 to 48 months, just like the SPI. The SPEI classification can be found in Table 2.

The classical approximation (Abramowitz and Stegun, 1965) of SPEI is as follows.

$$SPEI = W - \frac{C_0 + C_1W + C_2W^2}{1 + d_1W + d_2W^2 + d_3W^3} \tag{2}$$

where

$$W = -2\ln(p)$$

for $P \leq 0.5$, P being the probability of exceeding a determined D value, $P = 1 - F(x)$. If $P > 0.5$, P is replaced by $1 - P$ and the sign of the resultant SPEI is reversed.

The constants are $C_0 = 2.515517$, $C_1 = 0.802853$, $C_2 = 0.010328$, $d_1 = 1.432788$, $d_2 = 0.189269$, and $d_3 = 0.001308$.

Table 2: Wet and drought period classification according to the SPEI index.

SPEI Range	Class
2.0 and above	Extremely wet
1.5 to 1.99	Very wet
1.0 to 1.49	Moderately wet
−0.99 to 0.99	Near normal
−1.0 to −1.49	Moderately dry
−1.5 to −1.99	Severely dry
−2 and less	Extremely dry

Rainfall Anomaly Index (RAI)

The RAI was used to examine the frequency and intensity of dry and rainy years in the study area using daily rainfall data. Additionally, monthly RAI was calculated for specific years in the historical series to assess the distribution of rainfall during years with the greatest anomalies. Developed by Rooy (1965), RAI is calculated using the equation

$$RAI = \pm 3 \frac{P - \overline{P}}{\overline{E} - \overline{P}} \tag{3}$$

where $\overline{P}$ is measured precipitation, $\overline{E}$ is mean precipitation, and P is the average of 10 extrema and ±3 is the prefix used to limit the lower and upper bounds of the anomalies. Here extrema is meant by the highest and lowest rainfall years during the study period. According to the magnitude, RAI is classified as varying from extremely wet to extremely dry ranges, Van-Rooy (1965).

Table 3: Classification of Rainfall Anomaly Index.

RAI Range	Classification
Above 4	Extremely humid
2 to 4	Very humid
0 to 2	Humid
−2 to 0	Dry
−4 to −2	Very dry
Below −4	Extremely dry

Results and Discussion

Illustration for Meteorological Drought Indices

The study area, Kozhikode district in Kerala, India, is located in the humid tropical region and experiences poor distribution of rainfall with an annual average rainfall of above 3,000 mm. To analyze the dry periods, three meteorological drought indices—RAI, SPI, and SPEI—were adopted and compared. The results of the drought classification based on these indices for the given stations are presented in Table 4.

According to the SPI classification, the 12-month SPI showed the highest number of drought days (59) with the maximum number of extreme dry days (11). Severe drought days were observed in all months, and moderate drought days were more frequent in all months (3, 6, and 12).

For SPEI classification, the 3-month SPEI showed the highest number of drought days. The 12-month SPEI showed the maximum number of extreme dry days (10), with almost the same number of severe drought days in all months. The highest number of moderate drought days was observed in the 12-month SPEI of the Kottamparamba station.

In RAI classification, the Kottamparamba station showed the highest number of drought days (262). However, none of the stations showed any extreme dry days. The maximum number of dry days was observed in Kottamparamba (81), and the highest number of very dry days (181) was observed at the same station.

Table 4: Classification of drought based on SPI in, Kozhikode.

	SPI			
	Moderate drought	Severe drought	Extreme dry	Total
3-Month SPI	29	11	9	49
6-Month SPI	35	12	7	54
12-Month SPI	37	11	11	59
	SPEI			
3-Month SPI	188	14	6	208
6-Month SPI	185	11	6	202
12-Month SPI	184	9	10	203
	RAI			
	Extreme dry	Dry	Very dry	Total
RAI	0	81	181	262

Conclusion

The results of the study suggest that drought classification based on different indices is not uniform. While SPI identified 49 meteorological drought days, SPEI and RAI exhibited more, with 208 and 262, respectively. Since SPEI takes into account temperature information along with rainfall for index development, its results may be more reliable than those of the other considered meteorological drought indices.

R Codes for Drought Analysis

#Read in data

```
data <- read.table(file = “clipboard”, header = TRUE)
```

#Load necessary packages

```
library(precintcon) # for RAI analysis library(dplyr) # for grouping RAI results library(SPEI)
# for SPI and SPEI analysis
```

#Perform Rainfall Anomaly Index (RAI) analysis

```
rai1 <- rai(as.monthly(data), granularity = “m”) # monthly RAI based on monthly data
rai2 <- rai(as.monthly(data), granularity = “a”) # annual RAI based on monthly data
rai3 <- rai(as.daily(data), granularity = “m”) # monthly RAI based on daily data
rai4 <- rai(as.daily(data), granularity = “a”) # annual RAI based on daily data
```

#Export RAI outcomes

```
write.csv(rai1, file = “rai1.csv”)
```

#Plot RAI results

```
pplot.rai(rai1, granularity = “m”, xlab = “Month”, ylab = “RAI”, legend = NULL, fontsize
= 10, axis.text.color = “blue”, export = FALSE, export.name = “rai_plot.png”, width = 8.6,
height = 7.5, units = “cm”)
```

#Group RAI results

```
rai1$raic <- mutate(rai1, RAI = ifelse(rai < (-3), “extreme drought”, ifelse(rai < (-0.2),
“severe drought”, ifelse(rai < (-0.5), “moderate drought”, ifelse(rai < 0.33, “weak drought”,
“Normal”)))))
```

#Calculate Standardized Precipitation Index (SPI) per year on a daily or monthly precipitation series

```
spi1 <- spi(data, scale = 3) # SPI with a period of 3 months
```

#Export SPI outcomes

```
write.csv(spi1$fitted, file = “spi3.csv”)
```

#Plot SPI results

```
plot(spi1) plot(spi(ts(data$Rainfall, freq = 12, start = c(1991, 1)), scale = 6)) pplot.
spi(data$Rainfall, period = 3, distribution = “Gamma”, xlab = “Months”, ylab = “SPI”, ylim
= c(-3, 3), legend = NULL, fontsize = 10, axis.text.color = “black”, export = FALSE, export.
name = “spi_plot.png”, width = 8.6, height = 7.5, units = “cm”)
```

#Calculate Standardized Precipitation Evaporation Index (SPEI)

Here first calculate PET using Hargreaves method, then use that result to calculate SPEI

```
data$PET <- hargreaves(Tmin = data$TMIN, Tmax = data$TMAX, lat = 37.6475) # PET
spei3 <- spei(data$PRCP - data$PET, scale = 3) # SPEI with a period of 3 months
```

References

American Meteorological Society (AMS). Statement on Meteorological Drought. Bulletin of the American Meteorological Society 85, (2004): 771–3.

Aswathanarayana, U. 2001. Water Resources Management and the Environment. Rotterdam (Netherlands): Balkema.

Chang, F.C., Wallace, J.M. “Meteorological Conditions During Heat Waves and Droughts in the United States Great Plains.” Monthly Weather Review 115, no. 7 (1987): 1253–69.

Elzopy, K.A., Chaturvedi, A.K., Chandran, K.M., Gopinath, G., Surendran, U. “Trend Analysis of Long-Term Rainfall and Temperature Data for Ethiopia.” South African Geographical Journal 103, no. 3 (2021): 381–94.

Hasan, H.H., Mohd Razali, S.F., Muhammad, N.S., Ahmad, A. "Research trends of hydrological drought: A systematic review." Water 11 no. 11 (2019): 2252.

Lettenmaier, D.P., McCabe, G., Stakhiv, E.Z. 1996. Global Climate Change: Effects on Hydrologic Cycle. In L.W. Mays (Ed.), Water Resources Handbook, Part V. New York: McGraw-Hill. pp. 1–43.

Mishra, A.K., Singh, V.P. "A Review of Drought Concepts." Journal of Hydrology 391, no. 1–2 (2010): 202–16.

Riebsame, W.E., Changnon, S.A., Karl, T.R. 1991. Drought and Natural Resource Management in the United States: Impacts and Implications of the 1987–1989 Drought. Boulder, CO: Westview Press. p. 174.

Van Lanen, H.A.J., Peters, E. 2000. Definition, Effects and Assessment of Groundwater Droughts. In: Vogt JV, Somma F (Eds.), Drought and Drought Mitigation in Europe. Dordrecht: Springer. pp. 49–61.

Van Rooy, M. (1965). A Rainfall Anomaly Index Independent of Time and Space. Notes, 14, 6.

Webster, K.E., Kratz, T.M., Bowser, C.J., Adagnuson, J.J. The influence of landscape position on lake chemical responses to drought in Northern Wisconsin. Limnology and Oceanography 41, no. 5 (1996):977–84.

Wilhite, D.A. (2000). Drought as a natural hazard: concepts and definitions.

World Meteorological Organization (WMO). 1975. Drought and Agriculture. Technical Note No. 138, Report of the CAgM Working Group on Assessment of Drought. Geneva, Switzerland: WMO. p. 127.

6

Computation of Weather Indices for Climate Change Analysis

Venu Prasad H D[1] and Nandhana S[2]

[1]*Centre for Water Resources Development and Management, Kunnamangalam Kozhikode, Kerala*

[2]*Kerala University of Fisheries and Ocean Studies, Kerala*

Abstract

Nowadays, the impact of climate change is visible and we all are experiencing the same. However, the impact may vary from place to place and region to region. Extreme weather events like heavy downpour of rainfall, increase in temperature, etc., affect the population and the study of these extreme weather events help in resolving or combating the extreme events and planning for a better future. In this chapter, an attempt has been made to explain the extreme events, their types, and the analysis of events using two weather variables rainfall and temperature. The time series data of rainfall and temperature of Trivandrum Airport data were collected from the IMD station and the analysis of the same using ClimPACT software was shown as illustration. The results of the study conclude that the number of rainy days is increasing which indicates the probability of increasing rainfall and at the same time the temperature also shows an increase in trend in the station area.

Keywords: Climate change, climate indices ClimPACT, rainfall, temperature

Abbreviations Used

IPCC	*Intergovernmental Panel on Climate Change*
WMO	*World Meteorological Organization*
ETCCDI	*Expert Team on Climate Change Detection and Indices*
CCI	*Commission for Climatology*
WCRP	*World Climate Research Program*

JCOMM	*Joint Technical Commission for Oceanography and Marine Meteorology*
IMD	*India Meteorological Department*
ET-SCI Specific	*World Meteorological Organization's Expert Team on Sector Climate Indices*
CSDI	*Cold Spell Duration Indicator*

Introduction

In recent decades, severe weather and climatic occurrences have become more frequent, affecting human populations across the globe. These changes are caused, in part, by anthropogenic climate change. Vulnerable communities are particularly affected by these rapidly changing patterns, which can have catastrophic consequences. To mitigate the impact of climate change, communities must have access to accurate and up-to-date climate information. Weather predictions, from short-term forecasts to long-term scenarios based on data from previous decades and centuries, can provide valuable information to socioeconomic sectors that are vulnerable to climate change. Climate services offer scientifically-based data and forecasts that can assist decision-makers in managing the risks of climate variability and change. Globally accepted indices generated from daily temperature and precipitation data can indicate the more severe features of climate change. These indices are developed and analyzed by the CCI/WCRP/JCOMM Expert Team on Climate Change Detection and Indices (ETCCDI) and have made a significant contribution to the Intergovernmental Panel on Climate Change (IPCC) Assessment Reports.

What Is Climate Index?

The purpose of the index is to extract some useful information from a larger dataset; likewise, the climate index represents some particular information derived from a larger climate dataset. In broader terms, the climate index is the calculation that reflects a certain aspect of the climate record.

Why Do We Use Climate Indices?

Climate indices are vital tools that aid in our understanding of how climate patterns change over time, allowing us to better understand variability and trends in large datasets. When we rely solely on monthly averages of climate data, we risk missing crucial information that could have significant impacts on various sectors. In order to answer questions about climate features that affect human and natural systems, particularly with respect to extremes, indices derived from daily data are used to objectively extract information from weather observations. These indices can provide insights into the duration or magnitude of heat waves, the frequency and intensity of heavy downpours, and measurements of exceptionally wet, dry, hot, or cold periods with socioeconomic implications.

Basic statistics, such as mean, mode, and median, may not provide sufficient detail about specific events such as the frequency of warm or cold days, drought, or the number of wet or dry spells. For example, a 30-year dataset of daily maximum temperatures would contain approximately 10,950 daily values. By calculating the maximum daily temperature for each year, we can determine if the peaks of maximum temperature are changing over time, resulting in only 30 values, 1 for each year.

This index reduces the dataset to 30 values, providing insight into how the positive extremes of a specific climate record are evolving over time. Climate indices are thus a crucial tool for scientists and decision-makers to effectively manage the risks of climate variability and change, and to take necessary precautions to minimize their impacts.

There are different types of climate indices (Table 1) that serve different purposes.

One type of index is intensity indices, which describe the minimum or maximum of a variable at a set frequency. For example, the TXx index measures the highest daily maximum temperature during a given month. This index describes the intensity of the extreme event but does not provide information on how often it occurs or how long it lasts.

Frequency indices, on the other hand, measure the number of days above or below a certain threshold. For instance, the TX10p index measures the percentage of days in a month when the maximum temperature is below the 10th percentile. This index provides information on how often a particular threshold is crossed rather than on intensity or duration.

Duration indices are another type of index that measures the length of time of certain events or between certain events or the aggregate number of days meeting a specific criterion. For instance, the consecutive dry days (CDD) index describes the maximum length of a dry spell, measured by the maximum number of consecutive days with rainfall less than 1 mm (or any threshold). This index primarily describes the duration of the condition rather than the intensity or frequency of dry spells.

Finally, there are other specialized indices, such as the Standardized Precipitation Index (SPI), which provides a standardized distribution tool used to establish categories for very wet to very dry conditions. This index can describe the intensity, frequency, and duration of extreme precipitation events.

Table 1: Different Types of climate indices.

Category	Indices	ID	Definition	Units
Absolute and threshold	Minimum T_{min}	TNn	Annual minimum of daily minimum temperature (Tmin)	^{0}C
	Maximum T_{max}	TXx	Annual maximum of daily maximum temperature (Tmax)	^{0}C
	Frost days	FD	Days with Tmin < 0°C	Days
	Tropical night	TR	Days with Tmin > 20°C	Days
	Heavy precipitation	R10	Days when precipitation > 10 mm	Days
Duration	Consecutive dry days	CDD	Maximum number of consecutive days with total precipitation < 1 mm	Days
	Warm spell duration indicator	WSDI	Annual number of days contributing to events where 6 or more consecutive days experience TX > 90th percentile	Days
Percentile	Cool nights	TN10p	Total number of days when Tmin < 10th percentile	Days
	Cool days	TX10p	Total number of days when Tmax < 10th percentile	Days
	Warm nights	TN90p	Total number of days when Tmin > 90th percentile	Days
	Warm days	TX90p	Percentage of days when Tmax > 90th percentile	Days

The Expert Team on Sector-Specific Climate Indices (ET-SCI) is a team of climate scientists from around the world who are committed to enhancing the accessibility and consistency of sector-specific climate indices. They achieve this through the development of software, regional workshops, research, and training materials. One of the tools developed by this team is ClimPACT, a software package that enables the calculation of climate indices using daily temperature and rainfall data. These indices can be valuable for planning and decision-making in areas such as Disaster Risk Reduction, health, agriculture, and water management. ClimPACT is available both online and offline, making it accessible to users regardless of their connectivity. Sector specific climate indices from ClimPACT software

Index	Health	Agriculture and food security	Water resources and food security	Coasts	Disaster Risk Reduction	Energy	Fisheries	Forestry/GHGs	Cryosphere
FD Frost days	✓	✓			✓			✓	✓
TNlt2 TN below 2 °C		✓						✓	✓
TNltm2 TN below -2 °C		✓						✓	✓
TNltm20 TN below -20 °C		✓			✓			✓	✓
ID Ice Days		✓			✓			✓	✓
SU Summer days	✓				✓				
TR Tropical nights	✓	✓			✓			✓	
GSL Growing season length		✓			✓			✓	
TXx Max TX		✓				✓		✓	✓
TNn Min TN		✓				✓		✓	✓

Sectors considered in ClimPACT

- Agriculture and Industry
- Health
- Water resources
- Coasts
- Disaster risk reduction
- Energy
- Fisheries
- Forestry/GHGs

Demonstration of ClimPACT Software

Online ClimPACT (users can get the instructions from this link https://climpact-sci.org/)

ClimPACT is an R software package that calculates approximately 60 climate indices (Fig. 1) developed by the WMO Expert Team on Sector-Specific Climate Indices (ET-SCI). These are all based on maximum and minimum temperature and daily precipitation.

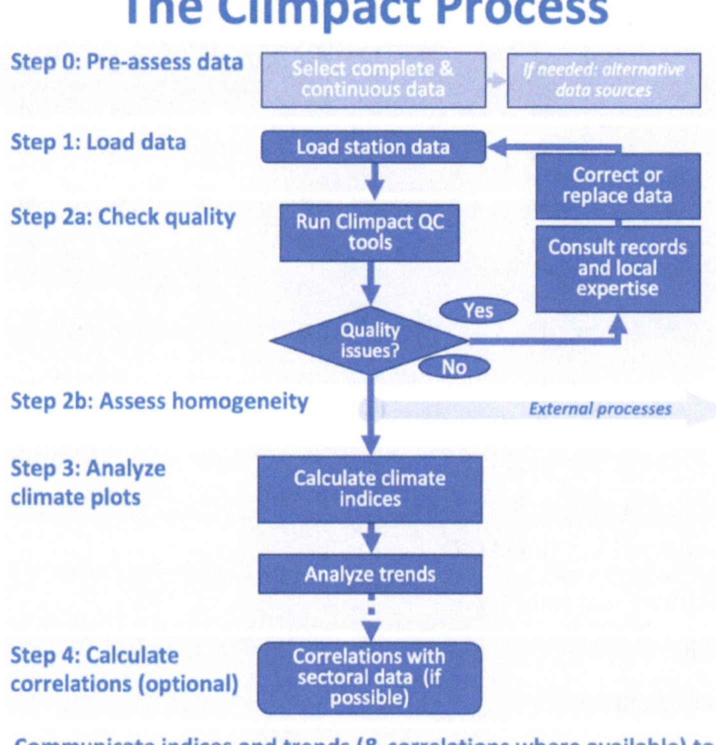

Fig 1: Flowchart of ClimPACT process.

Pre-assess Data

It is important to verify the quality and continuity of the data which is used for studying the climate indices. Poor data quality and missing data can affect the validity and accuracy of the results. It is impossible to have a dataset that is 100% complete therefore WMO guidelines can be used to look up whenever there is missing data.

Load Data

To compute climate indices, it is crucial to ensure the quality and homogeneity of the input data. The data must be in text format and include daily records of temperature and precipitation, as shown in the example dataset (Fig. 2) for Trivandrum airport collected from IMD station. The dataset should be checked for missing values and any gaps should be filled using appropriate statistical methods.

RF and TEM of TVM Airport - Notepad

File Edit Format View Help

YEAR	MN	DT	MAX	MIN	RF
1987	01	01	31.9	21.4	0
1987	01	02	31.6	21.7	0
1987	01	03	31.3	19.5	0
1987	01	04	30.7	19.4	0
1987	01	05	31.7	19.3	0
1987	01	06	31.5	20.4	0
1987	01	07	31.1	20.5	0.2
1987	01	08	31.1	20.2	0
1987	01	09	31.5	21.4	0
1987	01	10	31.6	23.8	0
1987	01	11	31.2	24.7	0
1987	01	12	31.6	22.8	0.4
1987	01	13	31.8	24.3	0.5
1987	01	14	31.9	21.7	0
1987	01	15	31.9	22.6	0
1987	01	16	31.8	22.3	0
1987	01	17	31.2	21.3	0
1987	01	18	31.8	21.7	0
1987	01	19	32.3	22.6	0
1987	01	20	31.9	21.2	0
1987	01	21	31.9	21.4	0
1987	01	22	31.7	20.9	0
1987	01	23	32.1	21.1	0
1987	01	24	31.6	19.2	0
1987	01	25	30.9	19	0
1987	01	26	31.2	19.9	0
1987	01	27	31.4	22.3	0
1987	01	28	31.9	21.9	0
1987	01	29	31.9	21.9	0
1987	01	30	31.9	22.4	0
1987	01	31	31.9	20.6	0
1987	02	01	31.6	21.4	0
1987	02	02	31.1	22.1	0
1987	02	03	31.2	21.4	0
1987	02	04	31.5	21.3	0
1987	02	05	31.6	20.2	0
1987	02	06	31.9	20.5	0
1987	02	07	31.5	20.5	0
1987	02	08	32.3	22.3	0
1987	02	09	31.6	22.3	0
1987	02	10	31.6	22.3	0

Fig 2: Trivandrum Airport station data.

Data Processing of Stations

The dataset is inputted in text format along with latitude and longitude. The base period, also known as the reference period, is used as a starting point to measure changes in climate variables. Ideally, observation data before and after the base period are needed. WMO recommends a standard base period of 1961–1990 for climate change analysis as it was considered the most stable period before the effects of global warming became evident. If this period is not available, the next recommended base period is 1981–2010. If neither of these periods is available, it is advised to choose the best and most common period among the stations used for analysis.

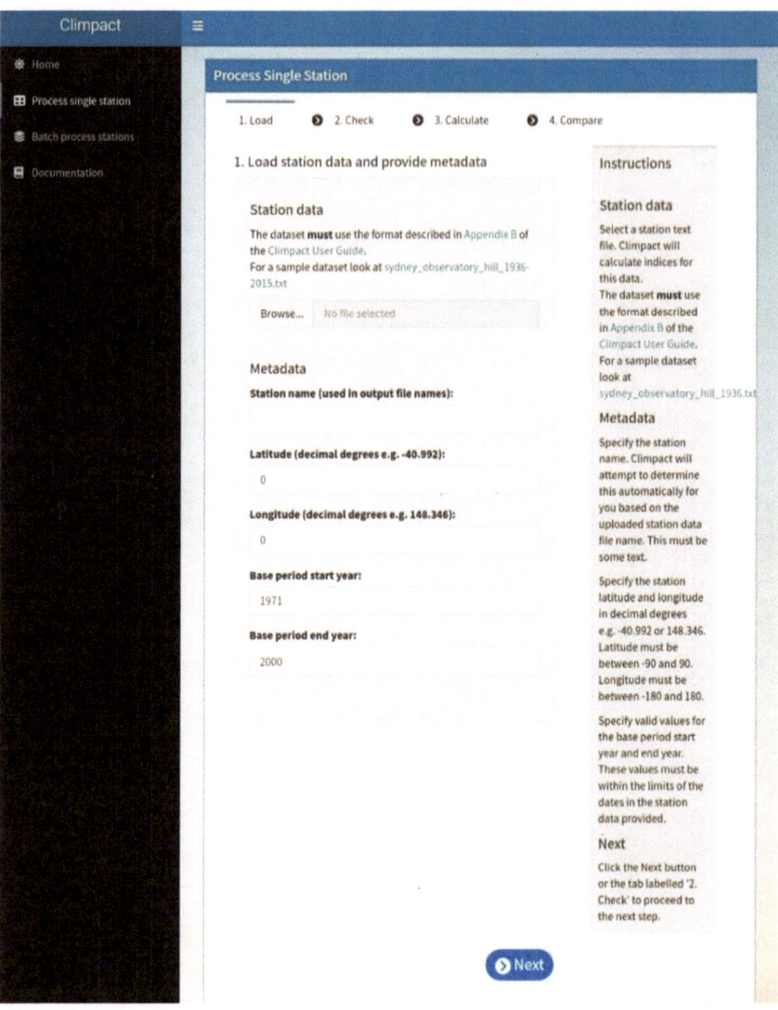

Fig 3: A screenshot of ClimPACT software.

Data Quality and Calculation of Indices

To generate accurate and useful climate indices, minimizing errors in the observed climate data is crucial. If quality issues persist after running quality control tools, it is necessary to investigate the reasons for these errors. Subsequently, the data undergoes homogeneity analysis and quality checking to remove outliers. Quality control files (Fig. 4) can be downloaded to aid this process.

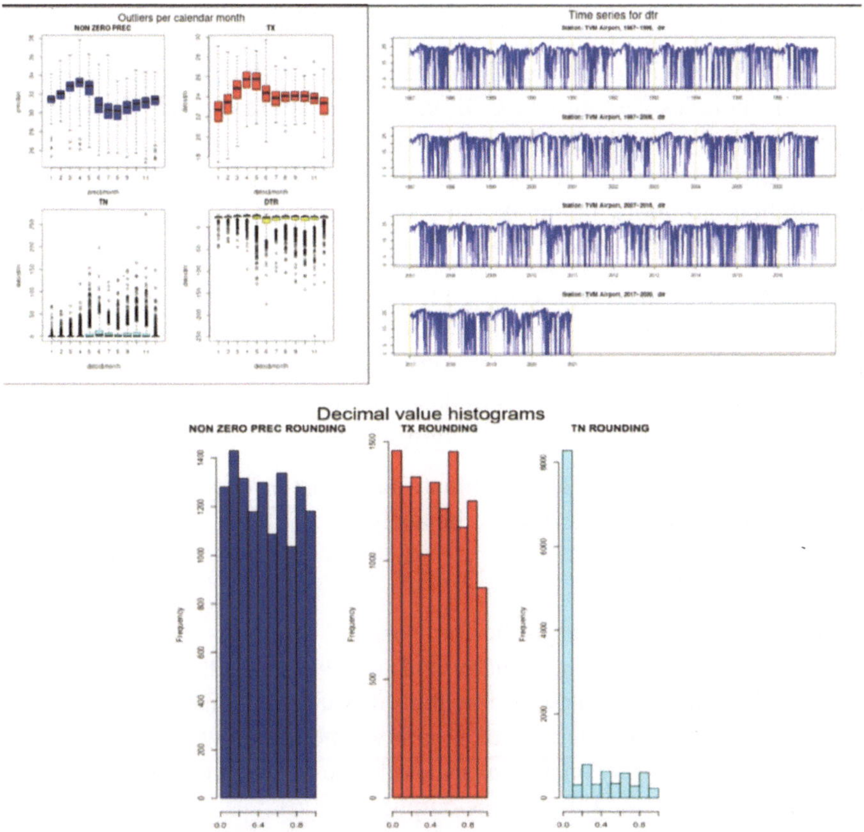

Outliers per calendar month

NON ZERO PREC

TX

TN

DTR

Time series for dtr

Decimal value histograms

NON ZERO PREC ROUNDING

TX ROUNDING

TN ROUNDING

Fig 4: Examples of ClimPACT quality control output.

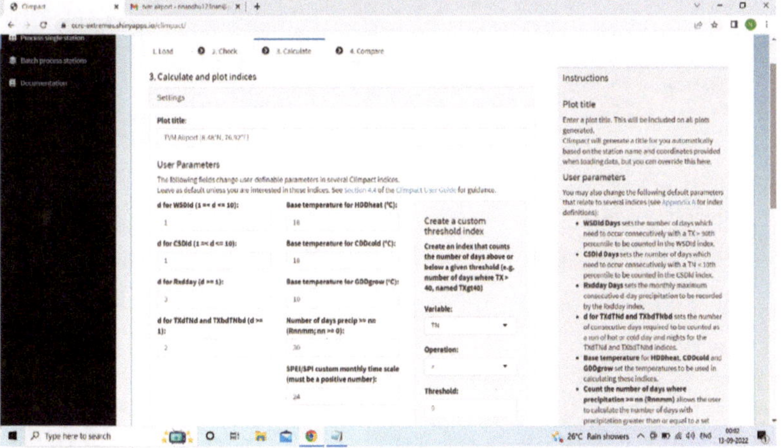

Fig 5: Calculate tab in ClimPACT.

Analyze Climate Plots: Calculation of Climate Indices

Climpact offers a diverse range of climate indices for different socio-economic sectors. It includes 27 core ETCCDI indices and approximately 34 additional indices created by ET-SCI. These indices cover a wide range, from basic indices such as maximum temperature, minimum temperature, and maximum precipitation, to more complex ones like frost days, cold spell duration indicator, and consecutive dry days.

To calculate these indices, users need to input the parameters for temperature and climate variables. The software provides default values, but users can also create threshold indices based on the study area, as temperature and rainfall can vary from station to station. A complete list of indices available in Climpact can be found at https://climpact-sci.org/indices/.

Prior to calculation, the observed climate data must be checked for quality and homogeneity to minimize errors in the analysis.

Results and Graphs

The final result is obtained in the form of .csv format as time series graphs and Excel data sheets both annually and monthly as well as graphical plots for every climate index calculated by the software. Some indices like RX1day are produced on annual as well as monthly time scales whereas indices like CSDI are produced only on annual scales. These results are then analyzed and conclusions are obtained from these outputs (Fig. 6).

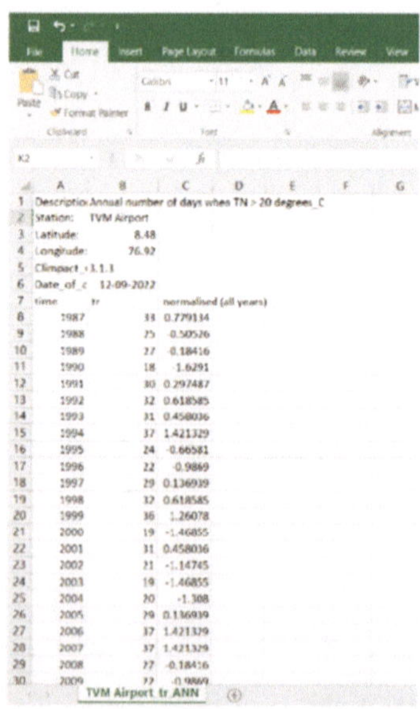

	A	B	C
1	Descriptio Annual number of days when TN > 20 degrees_C		
2	Station:	TVM Airport	
3	Latitude:	8.48	
4	Longitude:	76.92	
5	Climpact_v3.1.1		
6	Date_of_c	12-09-2022	
7	time	tr	normalised (all years)
8	1987	33	0.779134
9	1988	25	-0.50526
10	1989	27	-0.18416
11	1990	18	-1.6291
12	1991	30	0.297487
13	1992	32	0.618585
14	1993	31	0.458036
15	1994	37	1.421329
16	1995	24	-0.66581
17	1996	22	-0.9869
18	1997	29	0.136939
19	1998	32	0.618585
20	1999	36	1.26078
21	2000	19	-1.46855
22	2001	31	0.458036
23	2002	21	-1.14745
24	2003	19	-1.46855
25	2004	20	-1.308
26	2005	29	0.136939
27	2006	37	1.421329
28	2007	37	1.421329
29	2008	27	-0.18416
30	2009	22	-0.9869

Fig 6: A csv file showing the analyzed data.

Analyze Trends

A trend represents a pattern of change over time, such as an increase, decrease, or stability of values. It can be used to determine the rate of change in particular events, but it is not a forecast and cannot be used for long-term predictions. Climpact can generate trend lines using the Sens slope method, which provides three values for each plot: Sens slope, lower bound, and upper bound. The Sens slope value indicates the sign and magnitude of the trend, while the lower and upper bound are used to calculate the confidence interval, which indicates the accuracy of the trend. A smaller confidence interval indicates higher accuracy, and vice versa. To interpret the data, the Sen's slope value is considered if the p-value is significant.

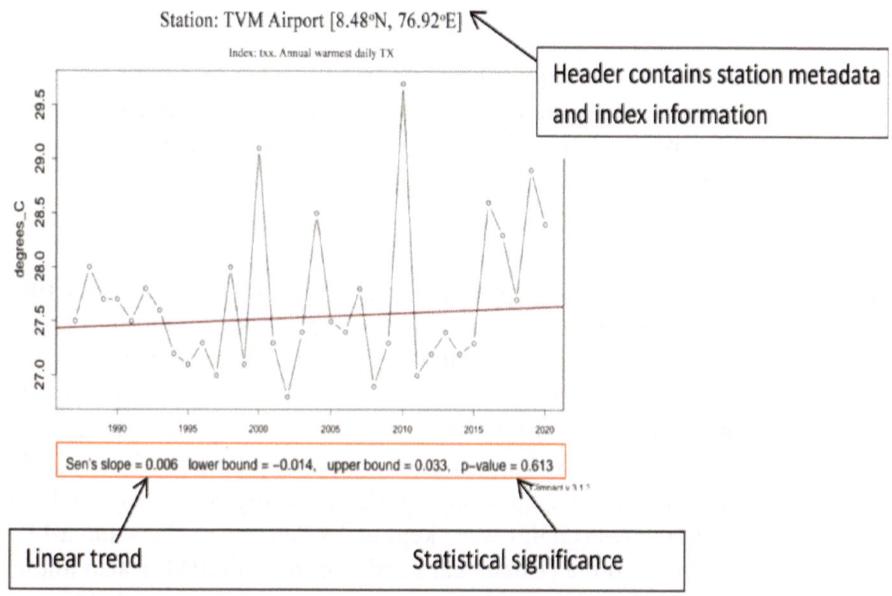

Fig 7: A ClimPACT generated graph.

The complete list of indices as well as guidelines on the description of indices is available: https://climpact-sci.org/indices/

Optional: Correlation with Sectoral Data

Climate indices can be correlated with sector-specific data to evaluate how changes in climate variables have affected a particular sector over a given period. By identifying patterns in climate change, climate data can provide accurate and reliable information that can be used to guide necessary adaptation actions. Climate indices are particularly important as they provide objective data that aid decision-making related to climate-related issues.

Some climate indices calculated by Climpact are included in IPCC AR4: https://www.ipcc.ch/report/ar4/wg1/observations-atmospheric-surface-and-climate-change/ .

Illustration Using Trivandrum Airport Data

Data Collection

Source of daily precipitation and temperature (both maximum and minimum) data from the rainfall observatory of Trivandrum airport station was acquired from the India Meteorological Department (IMD) data portal.

Studies of climate change require a long period of data in order to identify long-term trends rather than short-term fluctuations (WMO, 1996). Temperature and precipitation data was obtained for a period of 33 years from 1987 to 2020.

Climate Indices

A base period of 30 years was kept as a standard period which was from 1987 to 2017, while WMO recommended a base period of 1961–1990 for climate change analysis. This base period cannot be used as the collected data from the IMD station starts from 1987. Other than the default value provided by the software, 2 days were selected for WSDI and CSDI, 5 days for RXdays and TXnd, SPEI/SPI as 12, and the number of days of precipitation as 30 days. The base temperature for heating degree days (HDD), cooling degree days (CDD), and growing degree days (GDD) were kept at 18 degrees as the standard for heating and cooling in India (Bhatnagar *et al.*, 2018). ClimPACT also allows users to create threshold values for indices so indices such as TNle10, TNge25, TXle15, TXge35, PRge2.5, PRle2.5, and PRge64.5 were generated in the context of Kerala.

Results and Discussion

In the present study, spatial and temporal changes in temperature and precipitation indices defined by the Expert Team of Climate Change Detection and Indices (ETCCDI) are analyzed over Trivandrum Airport station.

Rainfall Indices

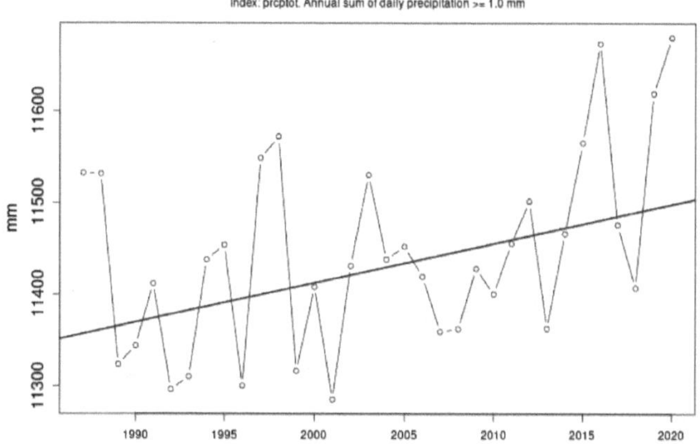

Fig 8: Graphical representation of PRCPTOT.

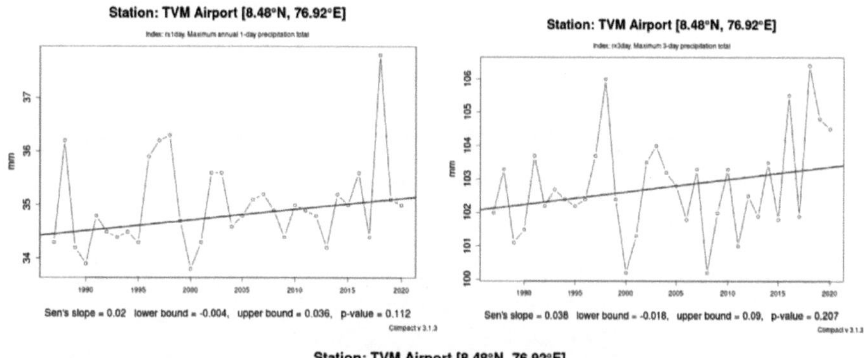

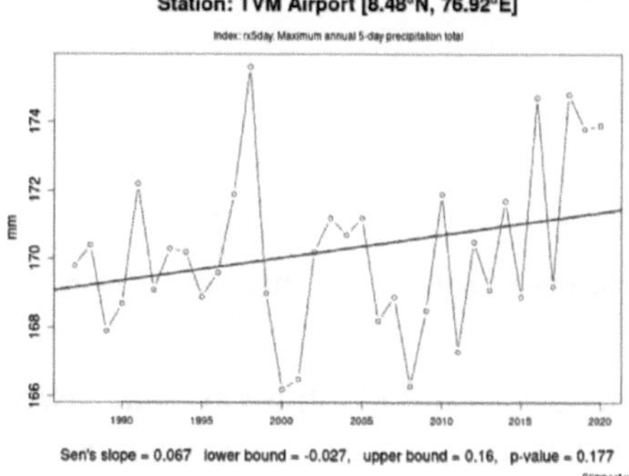

Fig 9: Graphical representation of RX1, RX3, and RX5 days.

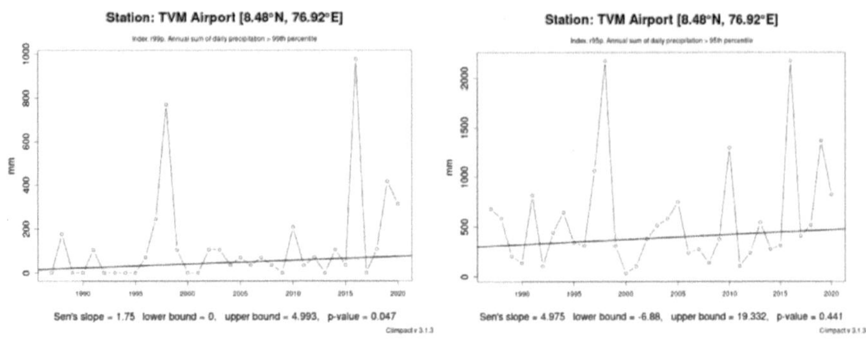

Fig 10: Graphical representataion of R99p and R95p.

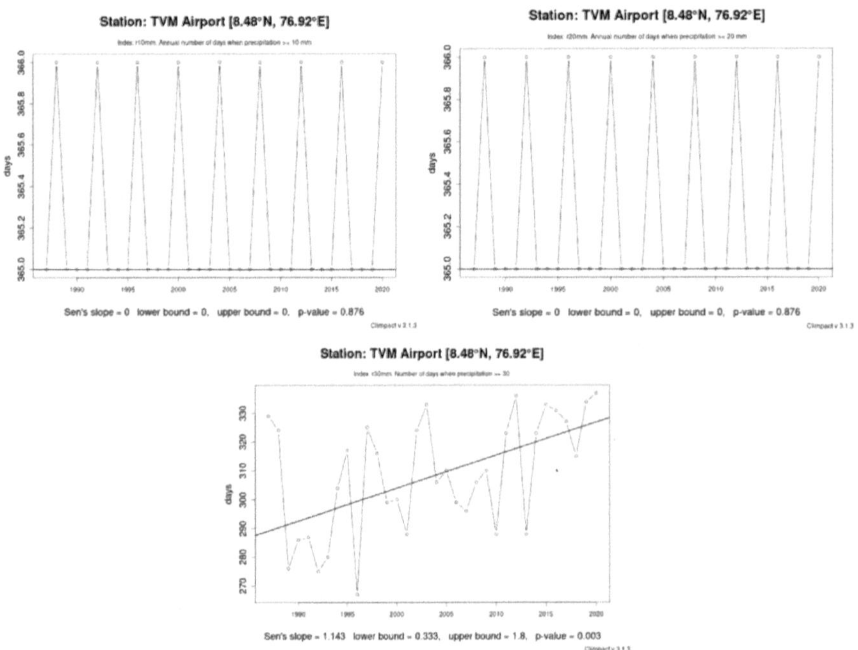

Fig 11: Graphical representation of R10mm, R20mm, and R30mm.

It is clear from the graphs that rainfall totals have increased at the station during the past 34 years, and that rainfall rates for successive one-day, three-day, and five-day periods are also rising. As the number of rainy days climbed, so did the overall number of days with more than 30 mm of rainfall. So, the probability of more frequent and intensified rainy days is foreseen in the study area.

1. Temperature Indices

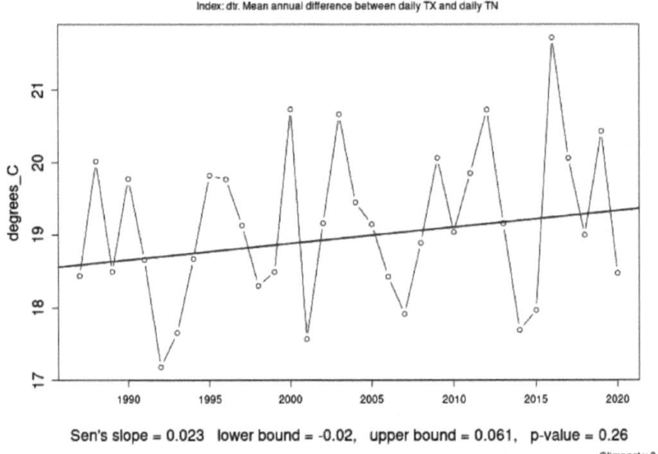

Fig 12: Graphical representation of DTR.

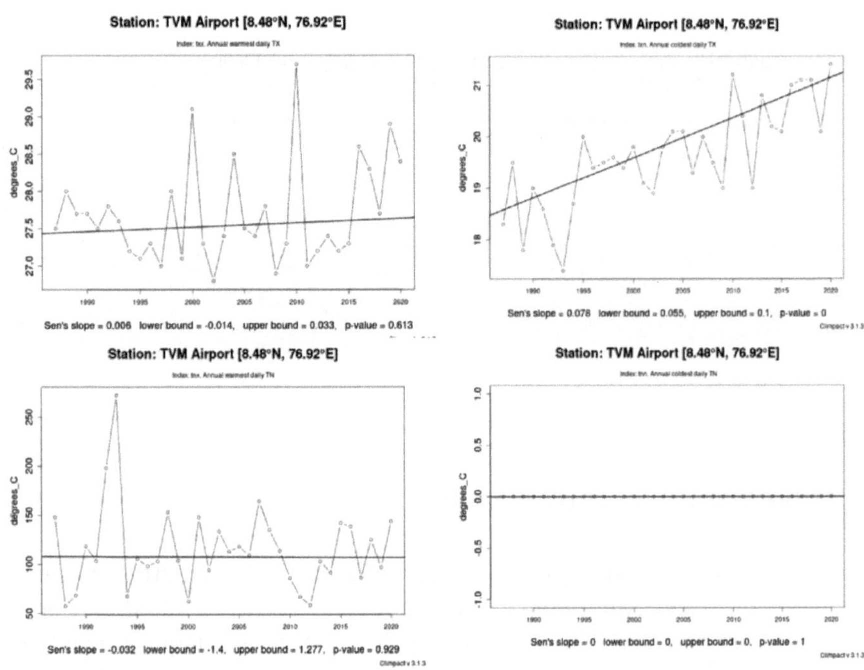

Fig 13: Graphical representation of TXx, TXn, TNx, and TNn

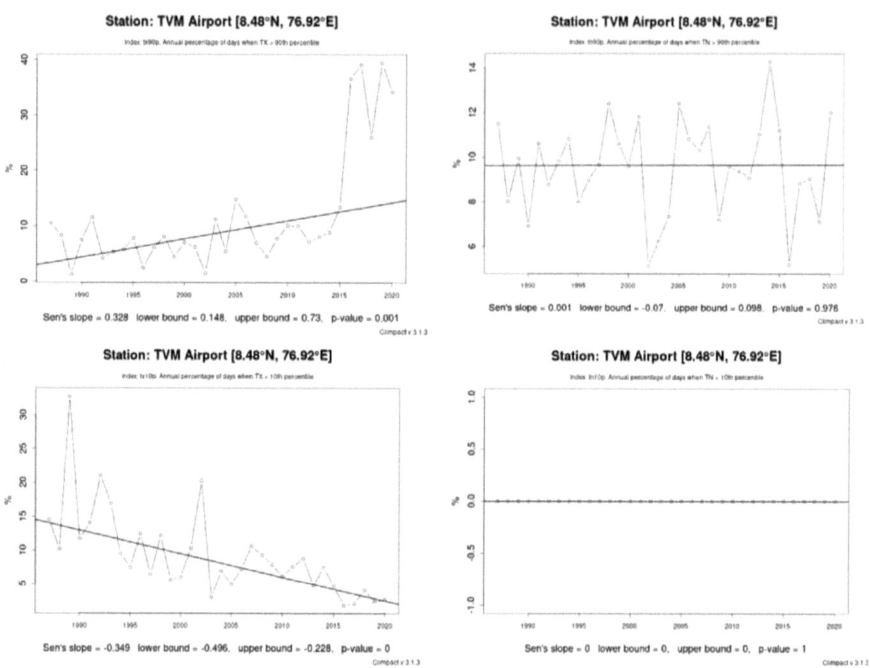

Fig 14: Graphical representation of TX10p, TX90p, TN10p, and TN90p.

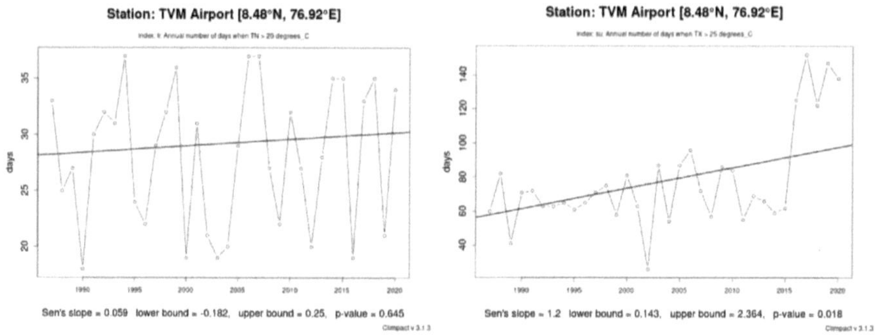

Fig 15: Graphical representation of SU and TR.

From the aforementioned graphs, it can be seen that there are fewer chilly evenings and more warm days and nights, as well as a rapid rise in the maximum temperature. The increase in index TXx shows unquestionably that regional temperatures are intensifying and becoming more susceptible to heat waves, droughts, and water shortages.

Conclusion

Thiruvananthapuram city is historically vulnerable to various natural disasters. The city is prone to naturally triggered hazards like floods, droughts, coastal erosions, lightning, and earthquakes and human-induced hazards like petrochemical accidents, transportation accidents, and epidemics (Kerala State Emergency Operations Centre, 2015). In the present study, spatial and temporal changes in the temperature and precipitation trends across Trivandrum Airport station were analyzed using a range of climate indices developed by ETCCDI, for example, the annual minimum of daily minimum temperature (TNn), annual maximum of daily maximum temperature (TXx), tropical nights (TR), cool nights (TN10p), cool days (TX10p), warm nights (TN90p), warm days (TX90p), annual total precipitation (PRCPTOT), consecutive dry days (CDD), heavy precipitation days index (R10), etc. The main findings are summarized as follows:

i. The diurnal temperature range is expanding as a result of the increase in the region's yearly maximum and minimum temperature. The increase in TX90p and drop in TX10p plainly indicate that colder days are declining and are being replaced by warmer days. This can cause cities to become urban heat islands, which produce more and more greenhouse gases and contribute to global climate change. The temperature in the area is rising, as evidenced by the rise in SU (when minimum temperature >20°C) and TR (when maximum temperature >25°C).

ii. The rainfall indices showed that the region's total rainfall is increasing (PRCPTOT) at a faster rate along with an increase in the maximum one-day, three-day, and five-day precipitation amounts. Additionally, more rainy days, including more extremely wet days, have been recorded. This shows that there are chances of more frequent and larger-scale heavy rainfall events in the studied region.

The rural-to-urban conversions in most populated cities like Trivandrum can increase the vulnerability of people to many hydrometeorological hazards. However, the situation will develop worse if the appropriate mitigation measures are not taken to tackle the worsening scenario. Therefore, climate change adaptive measures must be taken to gradual changes in average temperature, sea level, and precipitation; and mitigation and management must be done for more frequent, severe, and unpredictable extreme weather events.

References

Alexander, L., Herold, N. 2016. ClimPACT2. Indices and software. A Document Prepared on Behalf of The Commission for Climatology (CCl) Expert Team on Sector-Specific Climate Indices (ET-SCI). February, 46.

Bhatnagar, M., Mathur, J., Garg, V. (2018). "Determining Base Temperature for Heating and Cooling Degree-Days for India." Journal of Building Engineering 18 (2018): 270-80. https://doi.org/10.1016/j.jobe.2018.03.020

Kerala State Emergency Operations Centre. Hazard and Vulnerability Assessment Report - Thiruvananthapuram City, 2015.

Wazneh, H., Arain, M.A., Coulibaly, P. "Climate Indices to Characterize Climatic Changes Across Southern Canada." Meteorological Applications 27, no. 1 (2020): 1–19. https://doi.org/10.1002/met.1861

WMO.1996. Climatological Normals (CLINO) for the Period 1961–1990. WMO/OMM-No. 847, Geneva.

WMO. (2017). Guidelines on the Calculation of Climate Normals. WMO-No. 1203, 1203, 29. https://library.wmo.int/doc_num.php?explnum_id=4166

Section 3
Statistical Time Series Analysis for Water Resource Management

7

Data Uncertainties in Water Resources Modelling: A Special Reference to Hydrology

Dawn E S and Naveena K

[1]*Centre for Water Resources Development and Management, Kunnamangalam Kozhikode*

Abstract

Despite various progresses in understanding and representing the hydrological system, hydrological model outputs are still uncertain. The uncertainty arises from input data, calibration data, model structure, and the parameters involved. Recent climate changes have necessitated hydrological modeling studies incorporating future climate projections, which further add to the uncertainty. This chapter mainly discusses the data uncertainties associated with hydrologic modeling and the various analysis techniques that are in practice to decrease this. Dealing with missing observations in weather data is crucial to ensure accurate modeling results. This input uncertainty can be addressed by using imputation techniques. There are three main approaches to imputing missing values in weather data—univariate, multivariate, and spatiotemporal. Ultimately, the choice of imputation approach depends on the specific characteristics of the data and the research question being addressed. Univariate imputation may be sufficient for simple analyses, while multivariate or spatiotemporal imputation may be necessary for more complex analyses. It is important to carefully consider the advantages and disadvantages of each approach before selecting the most appropriate one for a given research question.

Keywords: Hydrological modeling, multivariate, random forest, uncertainties

Introduction

The hydrological cycle transfers water and energy between the land and atmosphere, but its regional characteristics make accurate modeling challenging. Hydrological models aim to predict hydrological events and processes, but there are uncertainties in their output due to the complex nature of the system and assumptions made. Empirical models use statistical or mathematical relationships, while process-based models use equations to represent physical processes.

Hydrological modeling involves various steps, leading to uncertainties that can affect the accuracy of model predictions. The four main types of uncertainties commonly encountered are data, model, parameter, and climate uncertainty. Data uncertainty arises due to errors or gaps in the input data, model uncertainty stems from limitations in the model structure or assumptions used, parameter uncertainty arises due to uncertainties in the values allocated to various model parameters, and climate uncertainty refers to uncertainties in the future climate projections used as inputs.

The input data required for hydrological modeling is obtained from a range of sources, including hydro-meteorological, catchment, and subsurface data. Despite various methods used for data acquisition and processing, the data available for modeling studies are often limited, with gaps and uncertainties. In many cases, the input data is derived from proxy variables or interpreted for the study region, leading to uncertainties ranging from 0% to 40% (McMillan *et al.*, 2018). Uncertainties in the input data can result in inaccurate water balance calculations and parameter estimations, which can affect the reliability of the model predictions. Several studies have highlighted the significance of input data uncertainty, especially in data-sparse regions (Engeland *et al.*, 2016; Bardossy *et al.*, 2020). Additionally, data uncertainty can also play a crucial role during the calibration process, as observed variables like streamflow and groundwater levels are often used, and these observations can be subject to similar uncertainties as the input data. Addressing data uncertainty is critical for improving the accuracy and reliability of hydrological modeling studies.

Accurate input data is crucial for reliable hydrological modeling, and input data uncertainty can arise from errors or gaps in precipitation data. Statistical techniques such as Monte Carlo (GLUE), Bayesian statistics, imputation techniques, and machine learning models can be used to study precipitation uncertainties. Missing observations in input data can significantly increase uncertainty in hydrological modeling, especially in stream flow prediction.

To address this issue, the Soil and Water Assessment Tool (SWAT) uses the ARCSWAT model, which employs a First Order Markov Chain algorithm for imputing missing data. However, it is essential to compare the effectiveness of this model with other imputation techniques to enhance the accuracy of hydrological predictions. Additionally, incorporating observation uncertainty can also impact parameter uncertainties and interactions between different uncertainties, as noted by Engeland *et al.*, (2016). Therefore, carefully addressing input data uncertainty and missing observations is crucial for improving the accuracy and reliability of hydrological modeling.

There are generally three common approaches to imputing missing values in weather data—univariate, multivariate, and spatiotemporal imputation techniques. Each approach has its own advantages and disadvantages depending on the nature of the data and the research question being addressed. Univariate imputation techniques rely solely on the historical data of the target variable to fill in missing values. The commonly used univariate imputation techniques for weather data include ARIMA, moving average, exponential smoothing, and artificial neural networks (ANN). This approach is simple to implement and requires minimal domain expertise, but may be biased if the missing values are related to other variables. Multivariate imputation techniques incorporate information from other relevant variables in addition to the target variable's historical data. The commonly used multivariate imputation techniques for weather data include ARIMAX, multiple imputations using chained equations (MICE), and random forest. This approach can be more accurate but requires more computational resources and domain expertise to select the appropriate set of predictor variables. Spatiotemporal imputation techniques are designed specifically for geographically distributed time-series data, such as weather data. They take into account both the temporal and spatial dependencies of the data to fill in missing values. The commonly used spatiotemporal imputation techniques for weather data include kriging, k-nearest neighbors (KNN), and Gaussian process regression (GPR). This approach can be very accurate but may require a large amount of data and computational resources.

In summary, the choice of imputation approach depends on the specific characteristics of the data and the research question being addressed. Univariate imputation may be sufficient for simple analyses, while multivariate or spatiotemporal imputation may be essential for more complex analyses. It is important to carefully consider the advantages and disadvantages of each approach before selecting the most appropriate one for a given research question.

R Code for Data Uncertainty Analysis: Multivariate Imputation Analysis

Load the data

```
k <- read.table(file = "clipboard", header = TRUE)
```

Method 1: Principle Component Analysis (PCA) for missing value imputation

```
install.packages("missMDA") library(missMDA)
```

Estimate the number of components using cross-validation

```
nb <- estim_ncpPCA(k, method.cv = "Kfold", verbose = FALSE)
```

Impute the missing values using PCA

```
res.comp <- imputePCA(k, ncp = nb$ncp, method = "Iterative")
```

View the completed dataset

```
View(res.comp$completeObs)
```

Method 2: Amelia algorithm for multivariate imputation

```
install.packages("Amelia") library(Amelia)
```

Impute the missing values using Amelia

```
res.amelia <- amelia(k, m = 100)
```

View the completed dataset

```
View(res.amelia$imputations$imp100)
```

Method 3: MICE algorithm for multivariate imputation

```
install.packages("mice") library(mice)
```

Impute the missing values using MICE

```
imp.mice <- mice(k, m = 100, method = "pmm")
```

Method 4: Random forest technique for multivariate imputation

```
install.packages("missForest") library(missForest)
```

Impute the missing values using missForest

```
k.imp <- missForest(k, verbose = TRUE)
```

View the completed dataset

```
View(k.imp$ximp)
```

Method 5: K-Nearest Neighbor model for spatio-temporal imputation of weather parameters

```
install.packages("VIM") library(VIM)
```

Impute the missing values using kNN

```
j <- kNN(h, variable = h$station2)
```

To compare the accuracy of the different imputation methods

We can use various evaluation metrics. The regr.eval() function from the DMwR package calculates several common metrics for regression models.

```
library(DMwR)
DMwR::regr.eval(p$Actual, p$PCA)
DMwR::regr.eval(p$Actual, p$Amelia)
DMwR::regr.eval(p$Actual, p$MICE)
DMwR::regr.eval(p$Actual, p$HMISC.Random)
DMwR::regr.eval(p$Actual, p$RandomForest)
```

Illustration for Uncertainty Analysis

Data Uncertainty: Minimization of input uncertainty by yhe adoption of advanced imputation techniques

Missing data can be a common issue in hydrological modeling as well, where the accuracy of the model outputs depends on the accuracy of the input data. Therefore, it is crucial to develop reliable imputation models that can accurately fill in the gaps in the data. In this study, we intend to compare the performance of five different multivariate imputation techniques suitable for time series hydrological parameters gap filling.

To do so, we used a 33-year dataset of maximum temperature data from January 1987 to March 2020. We evaluated the models' performance using accuracy measures like RMSE, MSE, MAE, and MAPE, considering the missed values in 2019 and 2020. The results indicated that Random Forest outperformed the other four models, namely HMISC, MICE, PCA, and Amelia algorithm (Table 1).

It is worth noting that the uncertainty in the input data can significantly affect the accuracy of hydrological modeling outputs. Therefore, it is essential to ensure the data's quality and completeness before using it in the modeling process. In cases where missing data cannot be avoided, developing accurate imputation models is crucial to mitigate the impact of the missing data on the model's accuracy. The findings of this study can provide valuable insights for researchers and practitioners working in the field of hydrological modeling and related disciplines.

Table 1: Accuracy evaluation of maximum temperature using RMSE, MAE, MAPE and RMSE.

Statistical Models	**MAE**	**MSE**	**RMSE**	**MAPE**	**Rank**
PCA	3.3	17.008	4.124	0.093	5
Amelia	3.525	17.828	4.222	0.1	4
MICE	2.52	10.61	3.257	0.071	3
HMISC	2.346	7.75	2.784	0.068	2
Random Forest	1.483	2.887	1.699	0.043	1

Summary

Hydrological modeling is vital for predicting hydrological events and processes, but uncertainties can impact the accuracy and reliability of model predictions. Input data uncertainty, caused by errors or gaps in data, can affect water balance calculations and parameter estimations, leading to unreliable predictions. Addressing input data uncertainty is thus crucial for improving modeling studies.

Imputation techniques can help address missing observations in input data and improve modeling accuracy. Univariate, multivariate, and spatiotemporal imputation techniques are commonly used, and each has its own strengths and weaknesses. The appropriate approach depends on the data and research question. Careful consideration of the pros and cons of each approach is necessary before selecting the best one.

References

Beven, K. "A manifesto for the equifinality thesis." Journal of Hydrology 320, no. 1–2 (2006): 18–36.

Engeland, K., Steinsland, I., Johansen, S.S., Petersen-Øverleir, A., Kolberg, S. "Effects of uncertainties in hydrological modelling. A case study of a mountainous catchment in Southern Norway." Journal of Hydrology 536, (2016): 147–160.

Her, Y., Yoo, SH., Cho, J., et al. "Uncertainty in Hydrological Analysis of Climate Change: Multi-Parameter vs. Multi-GCM Ensemble Predictions." Scientific Reports 9, (2019): 4974 https://doi.org/10.1038/s41598-019-41334-7

Joseph, J., Ghosh, S., Pathak, A., Sahai, A.K. "Hydrologic Impacts of Climate Change: Comparisons Between Hydrological Parameter Uncertainty and Climate Model Uncertainty". Journal of Hydrology 566, (2018): 1–22.

McMillan, H.K., Westerberg, I.K., Krueger, T. "Hydrological Data Uncertainty and Its Implications." Wiley Interdisciplinary Reviews: Water 5, no. 6 (2018): e1319.

McMillan, H.; Freer, J.; Pappenberger, F.; Krueger, T.; Clark, M. Impacts of uncertain river flow data on rainfall-runoff model calibration and discharge predictions. Hydrol. Process. 2010, 24, 1270–1284.

8

Trend Analysis For Studying Different Climate Change Scenarios Connecting To Water Resources

Naveena K, Surendran U, Drissia T K and K Ch V Naga Kumar

Centre for Water Resources Development and Management, Kunnamangalam Kozhikode

---

Abstract

Climate change poses a significant global threat that has far-reaching implications for both the environment and human welfare. As such, identifying trends in the hydrological cycle series is of paramount importance to mitigate its adverse effects. This chapter provides a comprehensive overview of various trend analysis methods for studying long-term moments in water-connected series, with a focus on the Modified Mann-Kendall (MMK) test, which is deemed superior to the traditional Mann-Kendall (MK) test. Our analysis of rainfall data from two gauging stations in Wayanad district, Kerala, revealed a significant downward trend in nearly all months, seasons, and annual periods, particularly in the post-monsoon season. Furthermore, a temporal variation study of streamflow time series data from 43 gauging stations in India revealed a decreasing trend in maximum flood values year by year in Northern Kerala, while central Kerala showed no significant trend. Conversely, the southern region, particularly stations located in dammed catchment areas, demonstrated a significant decreasing trend. These findings have critical implications for water management practices in the area and highlight the urgency of revising water management strategies for sustainable use of water resources. Additionally, our results illustrate that the MMK test is a superior alternative to the traditional MK test for trend analysis of auto-correlated series. Hence, we recommend adopting MMK models for trend analysis in such cases. In conclusion, accurately identifying and predicting trends in hydrological data is imperative, especially in the context of climate change. Our study underscores the urgency of implementing robust and effective trend analysis methods and addressing climate change's far-reaching impacts on the environment and human welfare.

Keywords: Climate change, Mann-Kendall (MK) test, modified Mann-Kendall (MMk) test, trend analysis

Introduction

Time Series Analysis (TSA)

TSA is a crucial tool in the study of climate change, as it allows researchers to carefully examine patterns and changes in weather parameters over time. By analyzing time series data for temperature, precipitation, and other weather variables, scientists can identify trends and develop models that accurately describe the underlying structure of the series. These models can then be used to make predictions about future changes in weather patterns and the potential impacts of climate change on the environment and human society. The accurate forecasting of climate change using TSA can inform policy decisions and facilitate adaptation strategies to mitigate the effects of a changing climate.

Components of Time Series

When analyzing a time series, we often want to separate the different components that contribute to its overall behavior. The classical decomposition method is a widely used technique that breaks down a time series into four components: trend, seasonal, cyclical, and irregular.

- The trend component $T(t)$ reflects the long-term movement of the series over time. In the context of climate change, this might be used to analyze the overall increase in global temperatures over the past century.
- The seasonal component $S(t)$ describes fluctuations within a year that are driven by weather and climate conditions. For example, the seasonal pattern of rainfall in a particular region may be analyzed to better understand the impact of climate change on local water resources.
- The cyclical component $C(t)$ captures medium-term changes in the series that occur in repeating cycles of two or more years. In the realm of climate science, this might be used to examine the cycles of El Nino and La Nina events and their impact on global weather patterns.

- The irregular component $I(t)$ captures random fluctuations that cannot be attributed to any of the other components. In the context of climate change, this might include the impact of natural disasters such as hurricanes or volcanic eruptions.

To model these components, two different approaches can be used: the multiplicative model and the additive model. The multiplicative model assumes that the components are not essentially independent and that they can affect one another, whereas the additive model assumes that the components are independent of each other. By understanding the different components of a time series and applying an appropriate model, we can gain valuable insights into the underlying patterns and trends that may be driven by climate change.

$$\text{Multiplicative model: } Y(t) = T(t) \times S(t) \times C(t) \times I(t)$$
$$\text{Additive model: } Y(t) = T(t) + S(t) + C(t) + I(t) \quad (1)$$

Climate Change and Climate Variability

Climate change refers to the long-term, continuous changes in average weather conditions that can lead to either an increase or decrease in temperature, precipitation patterns, and other environmental factors. In contrast, climate variability refers to shorter-term changes, such as those that occur over a month, a season, or a year. Given the inherent randomness and variability of weather patterns, it is important to study both long-term and short-term trends to gain a precise understanding of climate change. In the face of changing climatic scenarios, it is crucial to accurately identify weather movements in order to plan and implement various activities such as agricultural practices, cropping systems, and the mitigation of landslides and flood damages. Extreme weather events can have a significant influence on people's lives, making it important to accurately determine their potential trends.

To detect trends in weather parameters, numerous statistical methods have been developed. Non-parametric tests such as the Mann-Kendall (MK) test, Modified Man Kendal (MMK) test, and innovative trend analysis are commonly used for trend analysis using weather information. The linear regression method is also a parametric method for trend analysis, but its predefined assumptions can limit its usefulness for long-term moment analysis.

Mann-Kendall Test

The MK test is a nonparametric method for detecting long-term trends in time series data. This test calculates a test statistic using the sign of differences, which makes it less susceptible to nonlinearities compared to parametric tests such as linear regression (Sanjeevaiah *et al.*, 2021). The null hypothesis for the test is that there is no monotonic trend in the time series, while the alternative hypothesis is that there is a monotonic trend, either increasing or decreasing. The test statistic S is calculated using a simple formula.

$$S = \sum_{i=1}^{n-1} \sum_{j=i+1}^{n} sign(x_j - x_i) \tag{2}$$

where x_j and x_i are the annual values in years j and i, $j>i$ respectively, and n is the number of observations. The value of $sign(x_j - x_i)$ is computed as follows:

$$sign(x_j - x_i) = \begin{cases} 1 if (x_j - x_i) > 0 \\ 0 if (x_j - x_i) = 0 \\ -1 if (x_j - x_i) < 0 \end{cases} \tag{3}$$

For large samples ($n>10$), the test is conducted using a normal approximation with the mean ($E[S]$)and the variance ($Var(S)$)as follows:

$$E[S] = 0$$

$$Var(S) = \frac{1}{18}\left[n(n-1)(2n+5) - \sum_{p=1}^{q} t_p (t_p - 1)(2t_p + 5) \right] \tag{4}$$

Here q is the number of tied groups, and t_p is the number of data values in the pth group. The values of S and $Var(S)$ are used to compute the test statistic Z as follows:

$$Z = \begin{cases} \frac{S-1}{\sqrt{Var(S)}} & if\ S > 0 \\ 0 & if\ S = 0 \\ \frac{S+1}{\sqrt{Var(S)}} & if\ S < 0 \end{cases} \tag{5}$$

The presence of a statistically significant trend in the time series can be evaluated using the calculated Z value. A positive Z value indicates an upward trend in the series, while a negative Z value indicates a downward trend. The null hypothesis (H0) of no monotonic trend in the time series is rejected if the absolute value of the calculated Z value is greater than $Z_{1-\alpha/2}$, which is obtained from the standard normal cumulative distribution tables. The Z values are usually tested at a significance level of 0.05.

Wallis and Moore Phase-Frequency Test

The Wallis and Moore phase-frequency test is a statistical method used to test for independence in a series (Wallis & Moore, 1941). The test evaluates the hypothesis that the series is random in nature (H0) against the alternative hypothesis that the series is nonrandom in nature (Ha). The test statistic is calculated differently depending on the size of the series. For ordered series with n greater than 30, the test statistic can be expressed as:

$$Z = \frac{\left| h - \left(\frac{2n-7}{3} \right) \right|}{\sqrt{\frac{16n-29}{90}}} \tag{6}$$

Modified Man-Kendall Test (MMK)

The ML test is an extensively used statistical tool for detecting monotonic trends in time series data. However, its assumptions require that the observations are free from autocorrelation, as positive autocorrelation in the series can lead to misleading trend results (Yue *et al.*, 2002), increasing the likelihood of significant trends being detected. To address this issue, a MMK test was developed (Yue and Wang, 2004), which removes the linear trend component from the series and calculates the effective sample size using the lag-1 serial correlation coefficient. By correcting for autocorrelation, the MMK test eliminates the bias present in the original MK test, thereby improve its accuracy in detecting trends. Specifically, the variance of the MK test is replaced by the modified variance, and the remaining procedure follows the same steps to identify the trend. Overall, the MMK test is more accurate than the original MK test with respect to the empirical significance level, making it a useful tool for trend analysis in time series data.

The modified variance ($V(S)^*$) using effective samples size is given by:

$$V(S)^* = V(S) . \frac{n}{n^*} \tag{7}$$

Where n is the actual sample size (ASS), n/n^* is the correction factor (CF) and n^* is the effective sample size.

The effective sample size can be computed by:

$$n^{*} = \frac{n}{1 + 2\rho_k \sum_{k-1}^{n-1} \left(1 - \frac{k}{n}\right)} \tag{8}$$

Where ρ_k is the lag-k autocorrelation coefficient which can be estimated by sample autocorrelation for kth lag (ρ_k).

Sen's Slope Estimator

The Sen's nonparametric method is used to estimate the unit changes per year in the series. Here the trend in the series can be assumed to be linear.

$$f(t) = Qt + B \tag{9}$$

where B is a constant, t is time, and Q is the slope. The slopes of all data value pairs will be calculated to estimate the Q using the equation:

$$Q_i = \frac{x_j - x_k}{j - k} \tag{10}$$

where x_j and x_k are data values at time j and k ($j>k$), respectively. If there are n values x_j in the time series, there will be as many as $N = n(n-1)/2$ slope estimates Q_i. The Sen's estimator of slope is the median of these N values of Q_i. The N values of Q_i are ranked from the smallest to the largest and the Sen's estimator is

$$Q = Q_{\left[\frac{(N+1)}{2}\right]}, \text{ if N is odd or} \tag{11}$$

$$Q = \frac{1}{2}\left(Q_{\left[\frac{N}{2}\right]} + Q_{\left[\frac{(N+2)}{2}\right]}\right), \text{ if N is even.} \tag{12}$$

Code for Trend Analysis

```
# Load the required packages
library(trend)
library(modifiedmk)
# Read the data from a CSV file
data <- read.csv(file.choose(), header = TRUE)
#Apply Mann-Kendall Test for Trend
mk.test(data, alternative = "two.sided")
```

```
# Calculate the magnitude of trend using Sen's slope
sens.slope(data)
# Apply Seasonal Mann-Kendall Test for Trend
ts_data <- ts(data$Rainfall, start = c(1890,1), end = c(2008,12), frequency = 12)
smk.test(ts_data)
# Apply Wallis and Moore Phase-Frequency Test for Randomness
wm.test(data)
# Apply Modified Mankendal Test for Serial Correlation
tfpwmk(data)
mmkh(data)
```

Illustration

Illustration for Trend Analysis

Examination of Spatiotemporal Dynamics of Rainfall Pattern of Wayanad Region of Kerala

The Wayanad district of Kerala, located in the northern region of the state, is characterized by high altitudes and heavy rainfall. The livelihoods of a majority of the population in this region depend on agricultural activities. However, global climate change has caused significant shifts in rainfall patterns, making it essential to accurately identify potential trends in precipitation.

In this study, we focus on two gauging stations, namely Mananthavady and Vythiri, and analyze 33 years of annual rainfall data for all months and seasons, including the south-west monsoon (SWM) from June to September, north-east monsoon (NEM) from October to November, winter from December to February, and summer from March to May. By examining these data, we aim to increase a better understanding of rainfall patterns in Wayanad district and identify any potential trends that may impact agricultural activities and livelihoods in the region.

Table 1 presents the results of the statistical trend analysis of monthly, seasonal, and annual rainfall data (mm) in Wayanad district, Kerala, for the period of 33 years from 1986 to 2018. The randomness test using Wald-Wolfowitz Test showed that Manthawadi region had Z-statistic values greater than the Z-critical value (1.96) at a 5% level of significance for the months of January (2.83), February (1.98), March (1.98), and April (2.41). Similarly, Vettari showed significant dependency of the lag period for the months of January (3.26), February (2.40), and November (1.98), as their Z-statistic values were also greater than the Z-critical value of 1.96.

The MMK trend test is useful in identifying significant trends in the presence of autocorrelation. In the Wayanad region, the April month of the Mananthavady region and November and north-east Monsoon rainfall of Vettari region have shown significant downward trends in rainfall. In fact, all the months, seasons, and annual rainfall have shown a significant downward trend in Wayanad. According to the results of the Sens slope test, there is an average reduction of 14 mm of post-monsoon rainfall per year in Wayanad, with Mananthavady (−8.56 mm) and Vythiri (−10.94 mm) showing significant downward trends in Post monsoon season rainfall. Moreover, the number of rainy days (> 2.5 mm/day) during the last 33-year north-east rainfall has also shown a negative trend (Mananthavady (Z = −1.75), and Vythiri (Z = −1.42), with a reduction of about 63% in the number of rainy days during the last 10 years compared to previous years. This reduction in rainfall has a significant effect on crop productivity, especially coffee production in the Wayanad region, which is the predominant cash crop. The reduction in north-east monsoon increases the stress period of coffee crops, resulting in early maturity of the crop and irregular flowering. Establishing new clearings is difficult without supplementary irrigation in case of no north-east Monsoon. Additionally, the fear of Mealy bugs and sucking pest infestation also increases with a reduction in north-east monsoon.

Table 1: Rainfall trend analysis for Waynad region, Kerala.

	WM-test	**MK-test**	**MMK test**	**Sen's Slope**	**WM-test**	**MK-test**	**MMK test**	**Sen's Slope**
	Mananthavady				Vythiri			
Jan	2.83**	−1.17	−1.29	0.00	3.26**	−0.76	-0.88	0.00
Feb	1.98*	−1.19	−1.23	0.00	2.40*	−1.30	−1.16	0.00
Mar	1.98*	−0.11	0.10	0.00	1.56	0.23	0.27	0.074
Apr	**2.41****	**−1.54**	**−1.84**	−3.34	0.14	−1.18	−1.29	−4.92
May	1.13	0.07	−1.78	−0.25	0.57	−0.54	−0.367	−1.38
Jun	0.99	−0.29	−0.18	−2.9	0.99	−1.15	−0.76	−10.21
Jul	0.70	−0.33	−0.20	−2.75	1.84	−1.00	−0.90	-11.74
Aug	0.14	−0.29	−0.34	−2.04	0.71	−1.25	−1.08	−7.25

Sep	0.70	0.17	0.20	0.68	0.71	−0.31	−0.30	−2.30
Oct	1.84	−1.78	−1.67	−6.03	0.14	−1.60	−1.69	−7.72
Nov	0.15	−1.61	−1.64	−2.49	**1.98***	**−2.05***	**−2.13***	−3.71
Dec	0.70	−0.89	−0.97	−0.52	1.55	−1.56	−1.93	−3.25
SWM	0.28	0.01	0.05	0.00	0.71	−0.45	−0.37	−11.01
NEM	0.29	−2.06*	−1.98*	−8.56	1.84	−2.34*	−2.15*	−10.94
Winter	0.28	−1.14	−1.13	−0.96	**1.65**	**−1.71**	**−2.04***	−3.81
Summer	0.28	-0.99	−1.26	−0.28	0.14	−1.25	−1.23	−0.15
Annual	1.13	0.015	−0.01	0.01	0.71	−0.73	−0.623	−18.50

Trend Analysis of Streamflow in Kerala

The study analyzed daily discharge and flood series data from 43 gauging stations in India, obtained from the Central Water Commission and Water Resources Department. Different statistical tests were used to analyze the temporal variation in streamflow, including trend analysis. All tests were conducted at a 5% significance level, with a p-value threshold of 0.05. If the p-value was less than 0.05, the null hypothesis was rejected in favor of the alternate hypothesis (Drissia *et al.*, 2019). The trend analysis of flood series data from 43 gauging stations in Kerala, India, revealed that 16 stations showed a significant decreasing trend using the MK trend test, while the remaining stations showed no trend. Most of the stations with significant decreasing trends had data available from the 1960s. In Northern Kerala, the maximum flood values were decreasing year by year. In Central Kerala, only four stations showed a significant decreasing trend in flood series: Chaliyar, Kanjirapuzha, Karuvannur, and Ambalakkadavu. Among these stations, only Chaliyar, Kanjirapuzha, and Karuvannur showed a similar decreasing trend in daily streamflow. In contrast, Ambalakkadavu had a significantly decreasing flood series but a significantly increasing daily streamflow. Overall, most stations in Central Kerala showed a decreasing trend in flood series, except for a few toward the south. Some stations in South Kerala also exhibited a significant decreasing trend, such as Neeleswaram, Teekoy, Manimala, and Kallelli. Among these stations, Teekoy and Manimala also showed a significant decreasing trend in daily streamflow, but Kallelli had a significant increase in daily streamflow (Drissia *et al.*, 2019).

In the southern region, the gauging stations that showed a trend were mainly located in dammed catchment areas. Eight stations were found to have significant decreasing trends in both daily discharge and flood series: Shiriya u/s, Kakkadavu, Kannavam, Chaliyar, Kanjirapuzha, Karuvannur, Teekoy, and Manimala. These stations were not located nearby, except for Chaliyar and Kanjirapuzha, which belong to the same river basin, and Teekoy and Manimala, which belong to nearby river basins. Some stations had significant increases in daily discharge but significant decreases in flood series, such as Mangara, Ambalakkadavu, Neeleswaram, Kallelli, Punalur, and Pattazhi. Moreover, 21 stations showed significant trends in daily discharge but no significant trend in flood series. The researchers suggested that one of the reasons for these trends may be the presence of major or medium reservoirs in the catchment areas of many stations. If the inflows into these reservoirs were added, all stations might show the same trend (Fig. 1) in both daily and flood series (Drissia *et al.*, 2019).

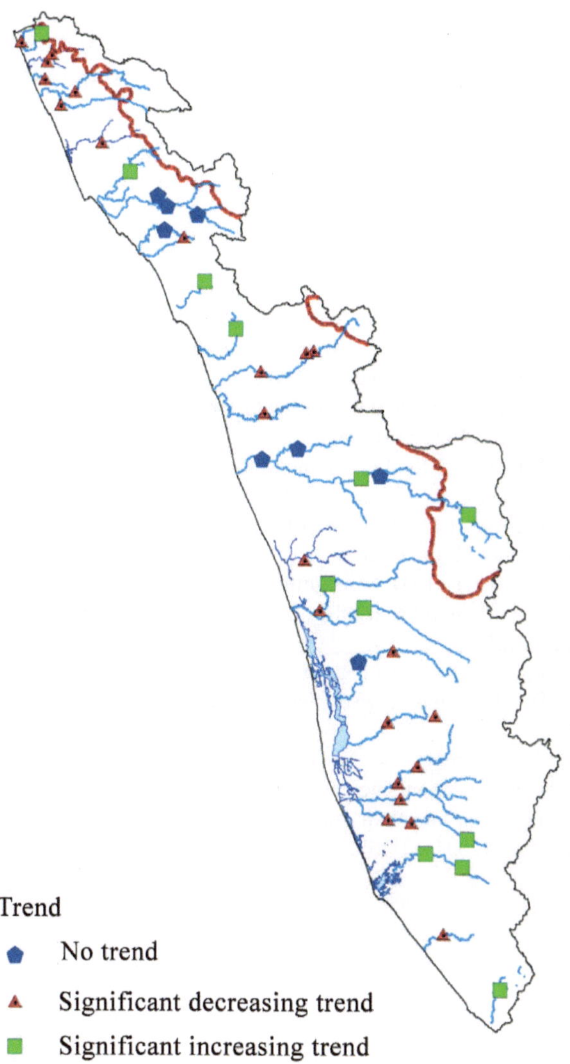

Fig. 1: Spatial variation of trend in Kerala.
Source: Drissia *et al.*, 2019

Conclusion

Due to the significant impact of recent changes in global climate on the distribution of weather series, it is crucial to accurately identify potential trends in these series. However, the presence of significant autocorrelation can suppress the results of the MK trend test. In such cases, the MMK test is known to outperform the standard MK test and provide more accurate trend analysis results. Therefore, the use of the MMK test is recommended for accurately identifying potential trends in weather series data affected by autocorrelation.

References

Drissia, T.K., Jothiprakash, V., Anitha, A.B. "Statistical classification of streamflow based on flow variability in west flowing rivers of Kerala, India." Theoretical and Applied Climatology 137, (2019): 1643–58. https://doi.org/10.1007/s00704-018-2677-0

Elzopy, K.A., Chaturvedi, A.K., Chandran, K.M., Gopinath, G., Surendran, U. "Trend analysis of long-term rainfall and temperature data for Ethiopia." South African Geographical Journal 103, no. 3 (2021):381–94.

Sanjeevaiah, S.H., Rudrappa, K.S., Lakshminarasappa, M.T., Huggi, L., Hanumanthaiah, M.M., Venkatappa, S.D., et al. (2021). "Understanding the Temporal Variability of Rainfall for Estimating Agro-Climatic Onset of Cropping Season over South Interior Karnataka India." Agronomy 11, no. 6 (2021): 1135.

Wallis, W.A., Moore, G.H. 1941. A significance test for time series and other ordered observations. In A Significance Test for Time Series and Other Ordered Observations (pp. 1–67). NBER.

Yue, S., Pilon, P., Phinney, B., Cavadias, G. "The Influence of Autocorrelation on the Ability to Detect Trend in Hydrological Series." Hydrological Processes 16, no. 9 (2002): 1807–29.

Yue, S., Pilon, P. "A Comparison of the Power of the t Test, Mann-Kendall and Bootstrap Tests for Trend Detection/Une comparaison de la puissance des tests t de Student, de Mann-Kendall et du bootstrap pour la détection de tendance." Hydrological Sciences Journal 49, no. 1 (2004): 21–37.

9

A Time Series Forecasting of Water Resources Using Autoregressive Integrated Moving Average (ARIMA) Models

Naveena K[1] and Santosha Rathod[2]

[1]*Centre for Water Resources Development and Management, Kunnamangalam Kozhikode*

[2]*ICAR-Indian Institute of Rice Research, Hyderabad*

Abstract

Autoregressive Integrated Moving Average (ARIMA) is one of the commonly used forecasting models in time series analysis. The popularity of the ARIMA model is due to its linear statistical properties as well as the popular Box-Jenkins methodology for model-building procedures. ARIMA models extract both seasonal and non-seasonal components in the data series by considering the present value of the series (r_t) *as a function of series-lagged values and stochastic error terms. The application of time series modeling in water resource management has broadened day by day to encompass the problems of global warming and climate change. Rainfall is one of the vital climatic factors that can indicate climate change. Spatial and temporal changes in rainfall would influence runoff, soil moisture, and groundwater reserves. Drastic changes in the rainfall lead to flooding and drought situations. In this connection, accurate prediction of future scenarios is very important to make decisions very early. Keeping this point into consideration, 30 years of monthly rainfall data of the humid tropical region, Kozhikode, Kerala, India was used to develop the best ARIMA model. After following proper identification, estimation and diagnostic checking between the ARIMA models were done. We observed the predicted values of rainfall match with observed values. So we can conclude that proper development of the ARIMA model by keeping importance to seasonality, Auto regression (AR), and Moving Average (MA) can increase the accuracy of the ARIMA model for short-term forecasts.*

Keywords: Autoregressive Integrated Moving Average (ARIMA), linearity, rainfall

Introduction

Time series analysis is a powerful tool for detecting and describing sequential processes, and it has become increasingly important in water resource management, especially in the context of global warming and climate change. Statistical methods have been widely used for analyzing climatological time series, such as precipitation, air temperature, and evapotranspiration, and hydrological time series, such as streamflow and surface water quality. However, most of the linear time series models depend on strict assumptions, such as homogeneity, trend-freeness, stationarity, non-periodicity, and lack of persistence, which are often violated in real-world datasets. Therefore, there is a need to develop new models that are more flexible and can accommodate different types of data (Machiwal & Jha, 2006).

One of the most popular and widely used classical time series models is the Autoregressive Integrated Moving Average (ARIMA) model, which has linear statistical properties and can be built using the Box-Jenkins methodology (Box and Jenkins, 1970). This model adoption is important for hydrological analysis because it can help in accurately characterizing and predicting the behavior of hydrological time series data. ARIMA models can capture complex patterns and trends in the data and can be used to forecast future values of the time series. This can help in better understanding the underlying processes that drive hydrological variability, such as changes in precipitation patterns or land use. Furthermore, hydrological data often exhibit seasonality and long-term trends, which can be accounted for in ARIMA modeling. ARIMA models can also be used for anomaly detection, identifying unusual events that may have significant impacts on water resources. Overall, ARIMA modeling provides a flexible and powerful framework for analyzing and forecasting hydrological time series, which is essential for effective water resource management in the face of changing environmental conditions.

Methodology

Stationarity Process of TS

Stationarity is a crucial assumption in time series analysis as it allows us to apply statistical methods that assume the data has constant mean, variance, and covariance over time. A strictly stationary time series has a joint distribution that is identical at all points in time, whereas a weakly stationary time series has a constant mean and covariance structure that is independent of time.

More specifically, {rt} is weakly stationary if:

1) E(rt) = μ, which is a constant, for all *t*.
2) Cov(rt , rt − s) = γs, which only depends on all-time *t* and lag *s*.

In contrast, a non-stationary time series does not have a constant mean and/or variance over time, making it difficult to apply traditional statistical techniques. To test for stationarity, the Dickey-Fuller test is a widely used statistical test. The test involves estimating a regression model, and if the AR parameter is close to zero, the original series may need to be differenced to achieve stationarity. If the Auto regression (AR) parameter is less than zero, the series is already stationary.

Autocorrelation Function (ACF)

The sample ACF is a fundamental tool for studying the dependence structure of a time series. The correlation coefficient examines the strength of linear dependence between two random variables *X* and *Y*, and takes values between −1 and 1. Assuming stationarity, the ACF ρ_k can be estimated by computing the sample correlation coefficient between pairs of observations that are k units apart in time. The lag-k autocorrelation coefficient is denoted by the symbol ρ_k and represents the correlation between Y_t and Y_{t-k}.

$$\rho_k = \frac{\sum_{t-k+1}^{T}(Y_t - \bar{Y})(Y_{t-k} - \bar{Y})}{\sum_{t-1}^{T}(Y_t - \bar{Y})^2} = \frac{\gamma_k}{\gamma_0};\ for\, k = 1, 2, \ldots\ldots\ where,\ \gamma_k = \text{cov}(Y_t, Y_{t-k})$$

Under the assumption of weak stationarity, the ACF ρ_k is defined as the covariance between Y_t and Y_{t-k} divided by the variance of Y_t. In other words, ρ_k measures the proportion of the total variance in the series that is described by the lag-k dependence between observations. By examining the sample ACF, we can identify the lag structure of the time series dependence and determine whether the series exhibits significant autocorrelation at various lags. This information is essential for choosing appropriate models and forecasting future values of the time series.

Partial Autocorrelation Function (PACF)

The PACF is a tool for studying the dependence structure of a time series that complements the ACF. While the ACF measures the relationship between observations at different lags, the PACF measures the correlation between observations at a specific lag after accounting for the correlation at shorter lags. More specifically, the PACF at lag k is the correlation between Y_t and Y_{t-k}, after removing the effects of the correlations at shorter lags (1, 2, ..., k−1). Mathematically, the PACF can be obtained from the autocovariance function γk as the solution to the following equation:

$$\phi_{kk} = \frac{p_k - \sum_{j=1}^{k-1} \phi_{k-1,j} p_{k-j}}{1 - \sum_{j-1}^{k-1} \phi_{k-1,j} p_j}, k = 2,3,.......where\ \phi_{k,j} = \phi_{k-1,j} - \phi_{k,k}\ \phi_{k-1,k-1}$$

Autoregressive (AR) Model

An AR model is a type of time series model that explains the current value of a time series variable as a linear function of its past values, along with a random error term. In the case of an equally spaced time series Y_t, with observations at time points t, t−1, t−2, and so on, the AR model can be represented as:

$$Y_t = \varnothing_1 Y_{t-1} + \varnothing_2 Y_{t-2} + \ldots + \varnothing_p Y_{t-p} + \varepsilon_t$$

Moving Average (MA) Model

The MA model is another widely used time series model that is useful for capturing dependencies in the data. In the MA(q) model, the value of the observed time series Y_t is a linear combination of q lagged error terms (ε_{t-1}, ε_{t-2}, ..., ε_{t-q}) and a random error term (ε_t) at time t. The equation for the MA(q) model can be written as:

$$Y_t = \varepsilon_t - \theta_1 \varepsilon_{t-1} - \theta_2 \varepsilon_{t-2} - \cdots - \theta_q \varepsilon_{t-q}$$

Autoregressive Moving Average (ARMA) Model

The ARMA model is defined as a linear combination of the past values of the time series data and the past error terms. It is represented as ARMA(p,q), where p represents the order of the autoregressive part, and q represents the order of the moving average part. The ARMA(p,q) model can be written as:

$$Y_t = \varnothing_1 Y_{t-1} + \varnothing_2 Y_{t-2} + \ldots + \varnothing_p Y_{t-p} + \varepsilon_t - \theta_1 \varepsilon_{t-1} - \theta_2 \varepsilon_{t-2} - \ldots - \theta_q \varepsilon_{t-q}$$

Autoregressive Integrated Moving Average (ARIMA) Model

ARIMA models can be classified into two main categories: pure models, which contain only AR or MA components, and mixed models, which contain both AR and MA components. The term “integration” (I) refers to the process of reverse differencing used to produce forecasts. An ARIMA model is typically represented as ARIMA(p, d, q), where p represents the number of lagged values used in the autoregressive model, d represents the order of differencing required to achieve stationarity, and q represents the number of lagged errors used in the moving average model (Box and Jenkins, 1994; Brockwell and Davis, 1996).

In general, the ARIMA model can be expressed as:

$$Y_t = \varnothing_1 Y_{t-1} + \varnothing_2 Y_{t-2} + \ldots + \varnothing_p Y_{t-p} + \varepsilon_t - \theta_1 \varepsilon_{t-1} - \theta_2 \varepsilon_{t-2} - \ldots - \theta_q \varepsilon_{t-q}$$

Y_t is the time series, $\varnothing_i$ and θ_j are model parameters, ε_t is random error, p is number of autoregressive terms, and q is number of moving terms

Seasonal Autoregressive Integrated Moving Average process (SARIMA)

SARIMA is one of the commonly using forecasting model and it includes seasonality component into ARIMA model. It will consider the present value of the series (r_t) as a function of series-lagged values and stochastic error terms. The seasonal ARIMA model is denoted by (p, d, q) (P, D, Q), where p denotes the number of autoregressive terms, q, the number of moving average terms, and d, the number of times a series must be differenced to induce stationarity. P, the number of seasonal autoregressive components, Q, the number of seasonal moving average terms, and D, the number of seasonal differences required to induce stationarity (Chang *et al.*, 2012; Naveena *et al.*, 2017). A seasonal autoregressive integrated moving average is expressed in the form:

SARIMA, which stands for Seasonal Autoregressive Integrated Moving Average, is a popular forecasting model that incorporates the seasonality component into the ARIMA model. It considers the current value of the time series as a function of its lagged values and stochastic error terms, taking into account both non-seasonal and seasonal patterns in the data. The SARIMA model is denoted by (p, d, q) (P, D, Q), where p denotes the number of autoregressive terms, q denotes the number of moving average terms, and d denotes the number of times the series must be differenced to induce stationarity. P denotes the number of seasonal autoregressive components, Q denotes the number of seasonal moving average terms, and D denotes the number of seasonal differences required to induce stationarity (Chang *et al.*, 2012; Naveena *et al.*, 2017).

The general form of a seasonal autoregressive integrated moving average model is expressed as:

ARIMA $(p, d, q) \times (P, D, Q)$ s

where s is the length of the seasonal cycle, such as 12 for monthly data with yearly seasonality.

$$\phi(B)\Phi(B)\nabla^d \nabla_S^D r_t = \theta(B)\Theta(B)\varepsilon_t$$

where

$w_t = \nabla^d \nabla_S^D r_t$ is number of regular differences.

$\nabla^d = (1 - B)^d$ is number of regular differences.

$\nabla_S^D = (1 - B^s)^D$ is number of seasonal differences.

The main stages in setting up a Box-Jenkins forecasting model are described below:

Identification

To identify the order of the AR and MA terms in an ARIMA model, one can follow a stepwise approach. First, it is important to check the stationarity of the time series. If the series is non-stationary, it needs to be differenced to achieve stationarity. Next, the ACF and PACF plots can be used to identify the initial order of the AR and MA terms. If the ACF plot shows a gradual decay pattern, then an MA model can be considered. If the PACF plot shows a gradual decay pattern, then an AR model can be considered. If both the ACF and PACF plots show gradual decay patterns, then an ARMA model can be considered. It is important to note that these initial orders are not always accurate, and multiple iterations may be necessary to arrive at the best order for the AR and MA terms.

Estimation of Parameters

During the estimation stage of the ARIMA modeling process, the coefficients of the identified models are estimated. This is typically done using the method of least squares, which involves minimizing the sum of squares due to residuals. The residuals represent the difference between the observed values and the values predicted by the model. The estimation process involves finding the values of the AR, MA, and seasonal parameters that result in the best fit to the data. The estimated coefficients are then used to construct the ARIMA model for forecasting future values of the time series.

Diagnostic Checking

After fitting a tentative ARIMA model to the data, it is essential to perform diagnostic checks to ensure that the model is adequate and potentially suggest improvements. One way to achieve this is through the analysis of residuals. A significant diagnostic check is the non-significance of autocorrelations of residuals, which can be evaluated via Portmanteau tests such as the Box-Pierce or Ljung-Box tests.

The Box-Pierce statistic, which is a function of the autocorrelations of residuals, measures the overall adequacy of the selected model. Its approximate distribution is Chi-square and is computed by summing the squares of the estimated autocorrelation at each lag up to a maximum lag of *k*. The Q statistic is then compared to critical values from the Chi-square distribution, with $(k - m - 1)$ degrees of freedom, where m−1 is the number of parameters estimated in the model. A small *Q* value with a large probability value indicates that the residuals are uncorrelated, and the model is correctly specified. In general, it is expressed as

$$Q = n \sum \mathrm{r}_2 \,(\mathrm{j})$$

Another modified *Q* statistic is the Ljung-Box statistic, which is given by a different formula. The formula for q includes the number of observations in the series, n, and is used to compare the *Q* statistic to critical values from the Chi-square distribution. If the *Q* or Ljung-Box statistic yields a significant value, it suggests that the chosen model does not fit well. A modified Q statistic is the Ljung-box statistic which is given by

$$Q = n\,(n+2) \sum r_2\,(j)/\,(n-j)$$

Forecasting

After the ARIMA model has been identified and estimated, it can be used to forecast future values of the time series. The model is applied to the future time points, and the forecasted values are generated. These predicted values can be compared to the actual values to calculate the forecast errors. This step is important in evaluating the performance of the model and its ability to accurately predict future values of the time series.

R Code for ARIMA model

Load required libraries

```
library(forecast)
library(tseries)
```

Import the data

```
ye <- read.table(file = "clipboard", header = TRUE)
```

Define the time series framework

```
ye.ts <- ts(ye, start = 1890, end = 2008, frequency = 1)
```

Plot the time series

```
plot.ts(ye.ts, col = 'blue', pch = 2, lwd = 3)
```

Plot the ACF and PACF

```
acf(ye.ts, lag.max = 40, col = 'red', pch = 2, lwd = 3) pacf(ye.ts, lag.max = 40, col = 'red',
pch = 2, lwd = 3)
```

Check for stationarity

```
stationarity <- adf.test(ye.ts) stationarity
```

Divide the data into training and testing sets

```
ye.training <- ye.ts[1:100] ye.testing <- ye.ts[101:119]
```

Fit an ARIMA model

```
arima1 <- arima(ye.training, order = c(1, 0, 1), include.mean = TRUE) arima1
```

Fit an ARIMA model using auto.arima function

```
ye.fit <- auto.arima(ye.training) ye.fit
```

Check for diagnostic test of residuals

```
res <- arima1$residuals res_test <- Box.test(res, lag = 1, type = c("Box-Pierce", "Ljung-
Box"), fitdf = 0) res_test
```

Calculate the accuracy of the model

```
accuracy(arima1)
```

Generate forecasts for the next 19 periods

```
fcast <- forecast(arima1, h = 19) fcast
```

Forecast the out-of-sample data

```
fitted.test1 <- data.frame(fcast) fitted.test <- fitted.test1[, 1] DMwR::regr.eval(ye.testing,
fitted.test)
```

Seasonal ARIMA r code

library(forecast)

 # Read in data

```
data <- read.csv("data.csv", header = TRUE)
```

```
# Convert to time series object
ts_data <- ts(data$Value, start = c(year(data$Date[1]), month(data$Date[1])), frequency = 12)
# Plot the time series
plot.ts(ts_data, main = "Time Series Plot")
# Check for seasonality using ACF and PACF
ggtsdisplay(ts_data)
# Fit a seasonal ARIMA model
model <- auto.arima(ts_data, seasonal = TRUE)
# Print the model summary
summary(model)
# Forecast future values
forecast <- forecast(model, h = 12)
# Plot the forecasted values
plot.forecast(forecast, main = "Forecasted Values")
```

Illustration: To Build the ARIMA

Model Using Monthly Rainfall Data of Humid Tropical Region

The monthly rainfall data of the humid tropical region, Kozhikode, India for the period of 30 years (1990–2020) was used to develop the ARIMA model. Data set from 1991 to 2019 was used for model development and 2020 data was used for validating the developed model.

Statistical Analysis of Rainfall

The statistical analysis of rainfall data is presented in Table 2. From the table, it can be seen that the mean annual rainfall is about 3129.4 mm. Looking at the amount of rainfall in different seasons (Table 2), the maximum rainfall received in the monsoon season followed by the summer season, and the minimum rainfall in the winter season. The coefficient of variation (CV) of the annual rainfall and seasonal rainfall varies between 55.6% and 19.6% indicating that there is significant variation in the rainfall between the seasons.

Seasonal Trends

The Mann–Kendall test, Sen's slope estimator, and the linear regression were also applied to detect the seasonal trends of precipitation (Tables 4 and 5). The increasing trend is 2371.03 mm/year for the south-west monsoon followed by 445.60 mm/year for the north-east monsoon, 228.77 mm/year for summer, and the Lowest trend obtained as 5.87 mm/year for winter.

Seasonal ARIMA

The Seasonal ARIMA Model possesses three stages of modeling namely, identification, estimation, and diagnostic checking. Identification was done after examining the ACF and the PACF (Fig. 1). The ACF plot and PACF plot in Figure 1a and b clearly indicate seasonality in the data set and nonstationary of the series. So, we went for seasonal differencing (D = 1), resulting in stationarity in the dataset (Fig. 1c and d). As per ACF and PACF plot results, we attempted different combinations of MA and AR orders [MA(0,1,2); AR(0,1); SMA (0,1,2,3); SAR(0,1,2,3)]. Parameter significance, minimum AIC and BIC criteria, and Box-Pierce test statistic results are used to identify the best combination of order and to check the goodness of the model. Among this ARIMA(0,0,1)(2,1,0) combination has shown all significant parameters, minimum AIC (4417.51) and BIC values (4432.72) as given in Table 3. The non-significant output of the Box-Pierce test statistic (chi-square =0.01) shows good fit of the model (Table 1). The obtained best model (ARIMA(0,0,1)(2,1,0)) was used to predict 12-month rainfall in advance and the results are given in Figure 2. Root mean square error (RMSE = 60.71) and Mean absolute error (MAE=37.25) values indicate a satisfactory level of prediction from the ARIMA(0,0,1)(2,1,0) model.

Table 2: Estimated parameters and Model fit statistics results for ARIMA (0,0,1)(2,1,0) model.

<table>
<tr><th></th><th>Estimate ± Std. Error</th><th>z Value</th><th>Pr(>|z|)</th><th>AIC</th><th>BIC</th><th>Box-Pierce Test Statistic</th></tr>
<tr><td>ma1</td><td>0.19±0.06</td><td>3.32</td><td><0.01**</td><td rowspan="3">4417.51</td><td rowspan="3">4432.78</td><td rowspan="3">0.01^{NS}</td></tr>
<tr><td>sar1</td><td>−0.74±0.05</td><td>−13.77</td><td>< 2.2e-16**</td></tr>
<tr><td>sar2</td><td>−0.25±0.05</td><td>−4.65</td><td>3.27e-06**</td></tr>
</table>

Notes: codes: 0 '***' 0.001 '**' 0.01 '*' 0.05 '.' 0.1 ' ' 1

Table 3: Predicted values and observed values for the year 2020 using ARIMA model.

Month	Predicted Values (mm)	Observed Values (mm)
January	0.0	0.0
February	0.098	0.0
March	29.52	20.0
April	76.41	66.6
May	282.61	191.6
June	776.48	642.2
July	930.18	802.7
August	614.24	618.8
September	267.25	285.7
October	183.41	206.3
November	82.26	106.3
December	52.91	57.8
RMSE	60.71	
MAE	37.25	

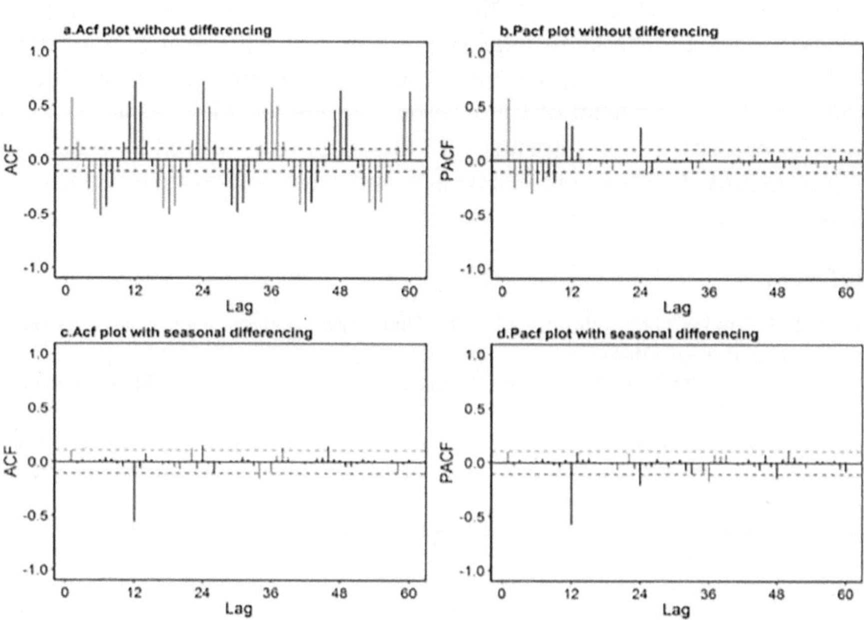

Fig 1: ACF and PACF plots at different lags of Rainfall data.

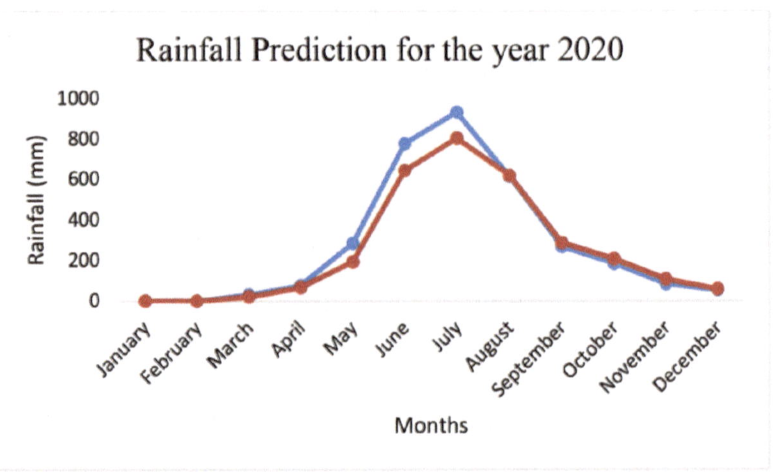

Fig 2: Rainfall predicted values and observed values for the year 2020.

Conclusion

The predicted values for the year 2020 using ARIMA model and validating the results with observed values. The findings show that the observed and projected values are nearer. The knowledge of the temporal pattern of rainfall trends analyzed in this study is a basic and important requirement for agricultural planning and management of water resources, and the ARIMA model is also good for predicting short-term analysis; this model can be used for prediction, helpful in providing warning in advance, and can take preventative measures in the district.

References

Box, G.E.P., Jenkins, G.M. Reinsel, G.C. 1994. Time Series Analysis: Forecasting and Control. Delhi: Pearson Education.

Chris Chatfield, C. 2003. The Analysis of Time Series: An Introduction, Sixth Edition. Chapman & Hall/CRC.

Machiwal, D., Jha, M.K. Time Series Analysis of Hydrologic Data for Water Resources Planning and Management: A Review. Journal of Hydrology and Hydromechanics 54, no. 3 (2006): 237–57.

Makridakis, S., Wheelwright, S.C., Hyndman, R.J. 1998. Forecasting: Methods and Applications. New York: John Wiley.

Naveena, K., Singh, S., Rathod, S., Singh, A. "Hybrid ARIMA-ANN Modelling for Forecasting the Price of Robusta Coffee in India." International Journal Current Microbiology Applied Science 6, no. 7 (2017): 1721–26.

Pankratz, A. 1983. Forecasting with Univariate Box-Jenkins Models: Concepts and Cases. New York: John Wiley & Sons.

10

Modeling Time-Varying Volatility of Drought Using ARIMA-GARCH

Naveena K[1], Halagundegowda G R[2] and Nagaraja M S[3]

[1]Centre for Water Resources Development and Management, Kunnamangalam Kozhikode

[2]Central Silk Board, Bengaluru

[3]Keladi Shivappa Nayaka University of Agricultural and Horticultural Sciences, Shivamogga

Abstract

This study focuses on the use of non-structural-mechanical time series models for modeling and forecasting drought. Specifically, the Box-Jenkins Autoregressive integrated moving average (ARIMA) and Generalized autoregressive conditional heteroscedastic (GARCH) models are examined. The results indicate that while the ARIMA model was not effective in capturing the volatility present in the data set, the GARCH model was able to successfully capture this volatility. To develop the ARIMA-GARCH model, a standard precipitation index (SPI) Drought index was created using monthly rainfall data from the humid tropical region of Kozhikode, India, over 30 years from 1991 to 2019. The Akaike Information Criterion (AIC) and Schwarz criteria were used to select an appropriate mean effect model and variance effect model. The ARIMA (0,0,1) model was selected as the mean model, as it exhibited a lower AIC (634.13) and Schwarz Criterion (645.68). For the variance effect model the GARCH (1, 2) model was found to have a lower AIC and Schwarz Criterion. Ultimately, the ARIMA (0,0,1)-GARCH (1,2) model was determined to be the best option for forecasting drought (SPEI) in the humid tropical region of Kerala.

Keywords: *ARIMA model, forecasting, GARCH model, SPEI index, stationarity unit root*

Introduction

Climate variability can have significant impacts on agricultural production, affecting both the demand and supply sides of farming. As climate change continues to negatively impact our ecosystem, food production, and human livelihood, this topic has gained widespread attention from researchers and academics worldwide. Given that agriculture is highly dependent on factors such as precipitation and temperature, it is considered one of the most climate-sensitive sectors (Sagar *et al.*, 2022).

Climatic data is complex and nonlinear, with significant fluctuation across time components, resulting in greater volatility. This means that stochastic climatic variables, including rainfall, temperature, relative humidity, and other weather parameters, often exhibit conditional heteroscedasticity, leading to irregular residual mean and variance across time series.

To mitigate risks and negative effects on agricultural production, it is essential to model climatic volatility, allowing for an understanding of causes, patterns, and impacts. Drought volatility is a complex and nonlinear phenomenon that has a profound impact on agricultural production, affecting both the demand and supply sides of farms (Wilhite, 2000). Climatic data, such as rainfall, temperature, and relative humidity, is inherently volatile, exhibiting conditional heteroscedasticity and irregular residual mean and variance across the time series. Modeling the volatility of droughts and other climatic variables can help to identify patterns, causes, and impacts, enabling producers to make informed decisions about production and investment under uncertain climatic conditions.

In hydrology, drought time series modeling is a popular method for characterizing droughts, with the linear autoregressive moving average (ARMA) model being a commonly used approach. Its ability to model the seasonal characteristics of hydrological variables with only a few parameters has contributed to its popularity. However, nonlinear time series modeling approaches, such as the generalized autoregressive conditional heteroscedasticity (GARCH) model, have also gained popularity due to their use in finance in investigating volatility and time-varying risks for assets, stock markets, and returns.

While the ARIMA model focuses on imposing a specific structure of the conditional mean of a stochastic process, the GARCH model focuses on imposing a specific structure of the conditional variance of the process. Despite these differences, the two models can be perfectly compatible with each other as the mean of the process can be modeled as ARIMA, and the variance can be modeled as GARCH. This enables both models to be used together to gain a comprehensive understanding of the stochastic process and accurately model droughts in hydrology (Halagundegowda *et al.*, 2019).

Materials and Methods

The GARCH Model

With the emergence of recent climate change scenarios, there has been a significant increase in volatility in weather datasets. In such cases, the GARCH model has proven to be effective for weather prediction, as it enables the modeling of volatility clusters. Studies by Panday *et al.* (2019) and Hwang *et al.* (2018) have shown the advantages of using the GARCH model in conjunction with the ARIMA model as a mean equation model to predict monthly rainfall series. Bollerslev (1986) built upon Engle's (1982) original work by developing a technique that allows the conditional variance to be dependent on previous own lags, similar to an ARMA process. The GARCH model consists of two equations, the mean equation and the variance equation, allowing for more accurate modeling of complex and nonlinear phenomena.

Mean Equation

The ARMA series is often used in the mean equation model, which is crucial for accurately estimating the variance of a time series. This equation is typically specified using various statistical methods, such as PACFs, ACFs, AIC, and BIC. Accurate specification of the mean equation is essential for obtaining reliable variance estimates.

One commonly used ARMA process is given by Equation (1), where Y_*t* represents the value of the time series at time *t*, θ_0 and θ_1 are coefficients, and **ε_*t*** and ε_(*t*-1) represent the error terms at times *t* and *t*−1, respectively. Equation (2) shows that **ε_t** is a normally distributed random variable with a mean of zero and a standard deviation of **σ_*t*** multiplied by the squared value of ϵ_*t*.

To estimate the variance of a time series, it is necessary to assume that ϵ_*t* is normally distributed with a mean of zero and a standard deviation of one and that **ε_*t*/φ_*t*−**1 is also normally distributed with a mean of zero and a standard deviation of **σ_*t***. By incorporating these assumptions into the ARMA process, it becomes possible to accurately model and predict the behavior of a time series over time.

The ARMA series is often used in the mean equation model, which is crucial for accurately estimating the variance of a time series. This equation is typically specified using various statistical methods, such as PACFs, ACFs, AIC, and

BIC. Accurate specification of the mean equation is essential for obtaining reliable variance estimates.

ARMA (1,1) process

$$Y_t = \theta_0 + \theta_1 Y_{t-1} + \varepsilon_t + \gamma_1 \varepsilon_{t-1} \quad (1)$$

$$\varepsilon_t = \sigma_t \epsilon^2_t \quad (2)$$

Here, we assume that $\epsilon_t \sim N(0,1)$ and $\varepsilon_t/\varphi_{t-1} \sim N(0, \sigma^2_t)$

where

φ_{t-1} is available information up to $t-1$ period

ε_t is disturbance term

σ_t is unconditionally standard normal variable

Variance Equation

The variance equation involves the conditional variance, which is expressed as a linear combination of the error term squared and its own lags. The GARCH (1, 1) process is the most widely used model specification for GARCH, and it is represented in the form of a mathematical equation.

$$\sigma^2_t = \alpha_0 + \alpha_1 \varepsilon^2_{t-1} + \beta_1 \sigma^2_{t-1} \quad (3)$$

The generalization of this model is GARCH (p, q) process

$$\sigma^2_t = \alpha_0 + \sum_{i=1}^{q} \alpha_i \varepsilon^2_{t-1} + \sum_{j=1}^{p} \beta_j \sigma^2_{t-1} \quad (4)$$

A sufficient condition for the conditional variance to be positive is

$$\alpha_0 > 0,\ \alpha_i \geq 0 \ \text{i=1,2....q};\ \beta_j \geq 0\ j = 1,2.......p.$$

The GARCH (p, q) model, which stands for generalized autoregressive conditional heteroskedasticity, is a more flexible model than the ARCH model as it allows for both autoregressive and moving average components in the heteroscedastic variance. This means that it can capture a wider range of volatility patterns in financial data. Interestingly, the first-order ARCH (1) model is simply a GARCH (0, 1) model when p is set to 0 and q is set to 1.

One of the key advantages of the GARCH model is its ability to provide a more parsimonious representation of high-order ARCH models, which can be much easier to identify and estimate. In terms of estimation, Maximum Likelihood functions are used to estimate the model parameters (Engle, 1982). To test for the presence of ARCH/GARCH effect, Lagrange's Multiplier test (Lama, 2015) can be employed.

R Code for GARCH Model Using Drought Data Set

```
# Install and load necessary packages
install.packages("quantmod")
library(quantmod)
install.packages("rugarch")
library(rugarch)
# Load data and convert to time series object
da1 <- read.table(file = "clipboard", header = TRUE)
data <- ts(da1$SPEI, start = c(1991, 1), end = c(2019, 12), frequency = 12)
# Plot time series data
ts.plot(data, main = "", xlab = "Period", ylab = "Drought index (SPEI)")
plot(da1$SPEI)
# Identification of ARIMA order and residual analysis
arima <- auto.arima(data)
res <- arima$residuals
res_test <- Box.test(res, type = c("Box-Pierce", "Ljung-Box")) # Diagnostic checking
# Fit ARIMA-GARCH model
model <- ugarchspec(mean.model = list(armaOrder = c(1, 1)), variance.model = list(model
= "sGARCH", garchOrder = c(1, 1)), distribution.model = "norm")
fit <- ugarchfit(data = data, spec = model)
# Plot model diagnostics
plot(fit)
# Generate forecast using the fitted model
forecast <- ugarchforecast(fitORspec = fit, n.ahead = 3)
# Plot the forecast
plot(forecast@forecast$seriesFor, main = "ARIMA-GARCH forecast", xlab = "Period",
ylab = "Drought index (SPEI)")
```

Illustration for ARMA-GARCH Model

Forecasting of Drought Scenario in Humid Tropical Region Using ARIMA-GARCH Model

In this study, we aimed to forecast the drought scenario in a humid tropical region, specifically in Kozhikode, India (Fig. 1). We developed a drought index called the Standard Precipitation Index (SPI) using monthly rainfall data spanning 30 years (1991–2019). To accomplish this, we employed an ARIMA-GARCH model using the R statistical package. The model was then utilized to predict rainfall for a three-month out-of-sample set. This approach can potentially aid in drought mitigation and water resource management in the region by providing early warning and preparedness measures.

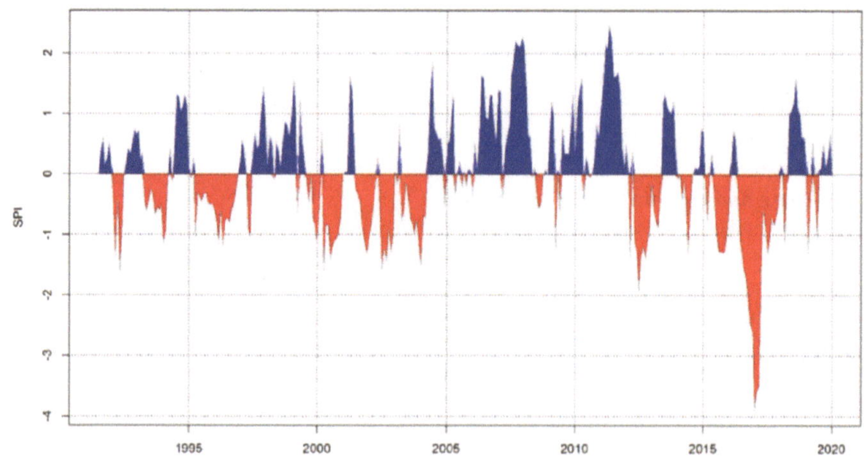

Fig 1: Time series plot SPEI drought index for humid tropical region (Kozhikode)

Figure 1 reveals that there are certain periods in which volatility is clustered, indicating the presence of an ARCH/GARCH effect in this series. This effect is further confirmed through the significance of the weighted Lagrange Multiplier (LM) test.

To select the most appropriate mean effect model and variance effect model for the available data sample, both the AIC and Schwarz criteria were utilized. The ARIMA (0,0,1) model was chosen as the mean model due to its lower AIC (634.13) and Schwarz Criterion (645.68). Among the various GARCH models considered, the GARCH (1,2) model was selected as the best model based on its lower AIC and Schwarz Criterion values.

The estimates of the GARCH (1,2) model indicate that the maximum coefficients of the mean and variance equations are statistically significant at the 5% level of significance. The non-significant test statistics of the Adjusted Pearson Goodness-of-Fit suggest that the model is a good fit for prediction. Therefore, we can conclude that the ARIMA (0,0,1)-GARCH (1, 2) model is the best among the various GARCH models for forecasting drought (SPEI) in humid tropical regions (Table 1).

Table 1: ARMA (0,0,1) , GARCH (1,2) model for drought prediction in Humid tropical region of Kerala (Kozhikode).

	Estimate	**Std. Error**	***t* Value**	**Pr(>|*t*|)**
Mu	−0.05	0.07	−0.70	0.49
ma1	0.65	0.05	14.33	0.00
omega	0.16	0.05	2.98	0.00
alpha1	0.32	0.09	3.41	0.00
beta1	0.38	0.21	1.83	0.07
beta2	0.00	0.24	0.00	1.00
Adjusted Pearson Goodness-of-Fit Test:			**Information Criteria**	
Group	**Statistic**	**ρ-value(g-1)**	Akaike	2.13
20	28.29	0.08	Bayes	2.20
30	31.32	0.35	Shibata	2.13
40	53.05	0.07	Hannan-Quinn	2.16
50	54.18	0.28		
3 Ahead Forecasting of Drought Using ARMA (0,0,1) , GARCH (1,2) model				
	Observed		**Predicted**	
Jan-2020	−0.29	Nearly Normal Drought	−0.05	Nearly Normal Drought
Feb-2020	−0.29	Nearly Normal Drought	−0.05	Nearly Normal Drought
March-2020	−2.87	Extreme Drought	−1.96	Severity Drought

Conclusion

The GARCH model outperforms the ARIMA model due to its ability to capture the time-varying conditional variance, which is essential for modeling volatility. In this study, we utilized the Standard Precipitation Index (SPI) Drought index, which was developed using monthly rainfall data of the humid tropical region of Kozhikode, India, spanning over 30 years (1991−2019), to develop an ARIMA-GARCH model. We also computed the AIC and the Schwarz Information Criterion (SIC), which indicated that the GARCH model had lower values than the ARIMA model. Based on these results, we can conclude that the GARCH model is a more appropriate model for forecasting drought.

References

Bollerslev, T. "Generalized Autoregressive Conditional Heteroskedasticity." Journal of Econometrics 31, no. 3 (1986): 307–27.

Engle, R. "Auto regressive Conditional Heteroskedasticity with Estimates of United Kingdom Inflation." Econometrica 50, (1982): 987–1008.

Halagundegowda G.R., Kantharaju B.M., Kumaresan P. "Time Series Forecasting of Price Volatility of Bivoltine Cocoons: An Application of GARCH Process and Artificial Neural Network." International Journal of Current Microbiology and Applied Sciences 8, no. 11 (2019): 840–50.

Hwang, Y.S., Kim, H.S.H., Yu, C. "The Empirical Test on the Impact of Climate Volatility on Tourism Demand: A Case of Japanese Tourists Visiting Korea." Sustainability 10, no. 10 (2018): 3569.

Lama, A., Jha, G.K., Paul, R.K., Gurung, B. (2015). "Modelling and Forecasting of Price Volatility: An Application of GARCH and EGARCH Models." Agricultural Economics Research Review 28, no. 1 (2015): 73–82.

Sagar, M., Mahadevaiah, G.S., Shripad, B., Harishkumar, H.V., Kiresur, V.R. Climate Variability and Its Impact on Cropping Pattern and Agricultural GDP in Central Dry Zone of Karnataka, India. Mousam 73, no. 2 (2022): 251–62.

Wilhite, D.A. 2000. Drought as a Natural Hazard: Concepts and Definitions. In D. Wilhite (Ed.), Drought: A Global Assessment, Vol. 1, pp. 3–18. London: Routledge.

Section 4

Machine Learning Approach for Water Resource Management

11

A Comparative Study of Machine Learning Techniques for Forecasting Effective Drought Index

Rajeev R K, Mrinmoy R, Kanchan S and Singh K N

ICAR-Indian Agricultural Statistics Research Institute, India

---

Abstract

A drought is a period when an area or region experiences below-normal precipitation. The lack of adequate precipitation can cause reduced groundwater, diminished stream flow, crop damage, and a general water shortage. Effective early drought warnings and predictions may be beneficial for the farmers. These statistical techniques are insufficient for providing long-term predictions. However, with improvements in machine learning techniques, there seems to be a mechanism to forecast drought in the long run with high accuracy. Frequently used machine learning algorithms in drought forecasting are artificial neural network (ANN), support vector machine, etc. Machine learning techniques have a powerful and versatile data-driven algorithm for capturing and representing complex input and output relationships. For illustration, we utilized monthly precipitation data for 52 years of Sagar and Chattarpur district of Bundelkhand region. The results confirm that the ANN model performs better compared to the SVR model in all the evaluation criteria.

Keywords: *Drought; Artificial neural network (ANN), forecast, support vector machine*

Introduction

Drought is an evolutionary process often triggered by precipitation deficiencies and may result in enormous economic losses. Effective early drought warnings and predictions may be beneficial for the farmers. To determine drought and its severity level, meteorological drought indices such as the standardized precipitation index, effective drought index etc. are utilized. Drought events can be foreknown using prediction models and by predicting these indices. Many studies on prediction have primarily focused on statistical approaches such as Holt-Winters, auto-regressive integrated moving average (ARIMA), etc. These statistical techniques are insufficient for providing long-term predictions. However, with improvements in machine learning techniques, there seems to be a mechanism to forecast drought in the long run with high accuracy. Frequently used machine learning algorithms in drought forecasting are artificial neural network (ANN), support vector machine, etc. Machine learning techniques have a powerful and versatile data-driven algorithm for capturing and representing complex input and output relationships.

Methodology

Artificial Neural Networks (ANNs)

ANNs have been used for modeling purposes in many disciplines like hydrology, meteorology, etc., for many years. Since its introduction in 1943 (McCulloch and Pitts, 1943), their popularity has increased primarily to address prediction issues involving stochastic variables, and nonlinear or unknown variations. As they are adaptable and less reliant on assumptions, there is no need to specify the underlying physical mechanism between inputs and outputs (Morid, *et al.*, 2007). This makes the ANN well-suited for drought forecasting in situations where the factors that cause a drought may not be completely known. Essentially, the ANN model learns how the input signal has changed over time based on historical data. It develops a logical input–output mapping system in order to forecast the future. Input data and associated output values are required to train and evaluate ANN models.

ANN is a computational paradigm made up of nonlinear components (neurons) functioning in parallel and massively interconnected via networks with varying weights. A neuron computes the sum of its inputs, adds a bias term, and pushes the result via a usually nonlinear activation function to generate a single output known as the neuron's activation level. ANN models are described by network architecture, neuron properties, and training or learning procedures (Lippman, 1987), with inputs, outputs, and hidden layers that are interconnected. The fundamentals' processing unit is a neuron, which computes a weighted sum of its input signals, y_i, for $i = 0,1,2,\ldots$, n, hidden layers, w_{ij} and then applies a nonlinear activation function to produce output signals u_j.

A neuronal model consists of an externally applied bias, b_k which has the effect of increasing or decreasing the net input of the activation functions depending on whether it is positive or negative. Mathematically, a neuron k may be described by

$$u_k = \sum_{j=1}^{m} w_k x_j$$

$$y_k = \Phi(u_k + b_k)$$

where $x_1, x_2, \ldots, x_m$ are the inputs signals; $w_{k1}, w_{k2}, \ldots, w_{km}$ are the synaptic weights of neuron k; u_k is the linear combiner output due to input signals; b_k is the bias; and $\Phi(.)$ is the activation function and y_k is the output signal of the neuron. Bias b_k has the effect of applying an affine transformation to the output u_k of the linear combiner in the model as shown by

$$v_k = u_k + b_k$$

In particular, depending on whether b_k is positive or negative, the relationship between the induced local field or activation potential v_k of neuron k and linear combiner output u_k can be modified. The bias b_k is an external parameter of artificial neuron k.

$$v_k = \sum_{j=0}^{m} w_{kj} x_j$$

$$y_k = \Phi(v_k)$$

The tangent sigmoid, $\phi(x)$, logarithmic sigmoid, $\Psi(x)$, and linear $\chi(x)$ transfer function are described as follows

$$\phi(x) = \frac{2}{1+e^{-2x}} - 1$$

$$\Psi(x) = \frac{1}{1+e^{-x}}$$

$$\chi(x) = linear(x) = x$$

Computationally efficient ANN networks employ second-order training methods, primarily the Levenberg-Marquardt (LM) or the Broyden-Fletcher-Goldfarb-Shanno (BFGS) quasi-Newton backpropagation learning algorithms that minimize the mean squared error between the predicted and observed variable (Tiwari and Adamowski, 2013). An LM algorithm uses an approximation to the Hessian matrix as follows

$$x_{k+1} = x_k - \left[J^T J + \mu I\right]^{-1} J^T e$$

where J is the Jacobian matrix calculated using standard backpropagation techniques and is less complex than computing the Hessian matrix. The J contains the first derivatives of network errors with respect to the weights and biases and **e** is a vector of errors. The BFGS quasi-Newton is an alternative to the conjugate gradient methods for fast optimization, which uses the following equation

$$x_{k+1} = x_k - A_k^{-1} g_k$$

where A_k^{-1} is the Hessian matrix (second derivatives) of the performance index at the current values of the weights and biases.

Least-Squares Support Vector Regression (LSSVR)

In this study, we incorporated the LSSVR model as a comparison with the neural network model. SVR-based models adhere to the structural risk minimization principle in contrast to neural network models which are developed to reduce empirical risk. SVR models are designed to improve the generalization properties of the forecast model by placing a regularization constraint on the size of the model weights. On the other hand, neural network models aim to reduce the empirical error. In general, the LSSVR algorithm (a specific model from the SVR class) is utilized to solve a linear programming problem where the training (input) vectors, taken as support vectors, are used to nonlinearly map inputs into forecasts. A linear formulation of LSSVR helps alleviate the convex quadratic programming problem associated with the original SVR model. In the standard SVR, the level of "slackness" (i.e., the property of the SVR which determines which training data vectors are chosen as support vectors) is set by selecting an inequality constraint. This forces the inclusion of another parameter in the model that must be optimized (i.e., via the quadratic programming problem mentioned earlier). However, the LSSVR model avoids this problem (and solves the regression problem as a set of linear equations) since each input vector is chosen as a support vector and a "slackness" term is not required (this aspect of LSSVR is often seen as an advantage over SVR due to its faster training speed, higher stability, and better control). Consequently, an LSSVR model provides global solutions to the error function which is an advantage over gradient-based algorithms (e.g., FFBP-ANN) that suffer from difficulties posed by the presence of local minima. A kernel function (K) and its relevant parameters are chosen such that a bound on Vapnik–Chervonenkis (VC) dimension is minimized, yielding stable solutions.

Given an input–output data set, X and Y comprised of N training data samples, i.e., (x_t, y_t) for $t = 1, 2, ..., N$, where $x_t \in \mathbb{R}^d$ and $y_t \in \mathbb{R}^d$; the LSSVR model is minimized with respect to the loss function L defined by

$$L(W,e) = \frac{1}{2}W^T W + C\frac{1}{2}\sum_{t=1}^{N} e_t^2,$$

where

e_t^2 is the quadratic loss term,

W is the weight vector and

C is the regularization parameter

To solve for the LSSVR parameters, the Lagrangian multipliers method is used as follows (Suykens *et al.*, 2002):

$$L(W,k,e,\alpha) = L(W,e) - \sum_{t=1}^{N} \alpha_t \left\{ W^T \phi(x_t) + k + e_t - y_t \right\}$$

where

$\phi(x)$ is a nonlinear mapping function

$\alpha \in R^N$ is the set of Lag range multipliers and

$k \in \mathrm{R}$ is the bias term.

The conditions which prove to be optimal in solving the LSSVR parameters are determined by taking partial derivatives of the extended loss function [i.e., L($W, k, e\alpha$)] with respect to each term ($W, k, e\alpha$) as follows:

$$\frac{\delta L}{\delta W} = 0 \rightarrow W = \sum_{t=1}^{N} \alpha_t \phi(x_t)$$

$$\frac{\delta L}{\delta k} = 0 \rightarrow \sum_{t=1}^{N} \alpha_t = 0$$

$$\frac{\delta L}{\delta e_t} = 0 \rightarrow \alpha_t = Ce_t, \quad t = 1, \ldots, N$$

$$\frac{\delta L}{\delta \alpha_t} = W^T \phi(x_t) + k + e_t - y_t, \quad t = 1, \ldots, N$$

The above conditions can be expressed in matrix form as:

$$\begin{bmatrix} 0 & \vec{1}^T \\ \vec{1} & \Omega + C^{-1}I \end{bmatrix} = \begin{bmatrix} K \\ \alpha \end{bmatrix} = \begin{bmatrix} 0 \\ Y \end{bmatrix}$$

where

$\vec{1}$ is a vector of ones and

Ω is used to represent the kernel function (K) satisfying Mercer's theorem:

$$\Omega_{uv} = \phi(x_u)^T \phi(x_v) = K(x_u, x_v),\ u, v = 1, \ldots, N,$$

where the kernel function $\left[K(x_u, x_v) \epsilon \mathbb{R}\right]$ may be represented by the commonly used RBF:

$$K(x_t, x) = e^{\left(-\frac{\|x - x^2\|}{2\sigma^2}\right)}$$

where σ represents the RBF kernel width.

Effective Drought Index (EDI)

In order to develop drought forecasting models, the precipitation data were used to calculate the EDI (Byun and Wilhite 1999).

Suppose that P_m is the rainfall recorded on any day, m $(1 \le m \le 365)$ and N^* is the duration of summation of the preceding period, then P_E for that (current i[th]) day is:

$$P_{E_i} = \sum_{N^*=1}^{D} \left[\frac{\sum_{m=1}^{N^*} P_m}{N^*} \right]$$

Above equation defines the degree to which Pm is converted into P_m for the ith day but importantly, this model considers antecedent precipitation with reduced weighting. That is, the P_E will accumulate 100% of precipitation received a day before, and 85% of that received 2 days before, and 77% of that received 3 days before, and so on, to 0.0423% of precipitation 365 days before the ith day (Kim *et al.*, 2009). The model puts the highest weight on present rainfall, whereas the previous days' contributions decrease gradually up to the annual cycle (N = 365 days, ignoring leap year for simplicity). In order to formulate the EDI, two simple equations were utilized:

$$DP_{E_i} = P_{E_i} - MP_E$$

$$EDI_i = \frac{DP_{E_i}}{SP_E}$$

where, MP_E is the mean P_E, SP_E is the standard deviation derived from the hydrological period (1951–2002), and *i* is the particular day. The "drought range" of EDI are:

Extremely dry conditions <−2

Severe drought at −1.5 to −1.99

Moderate drought at −1 to −1.49.

Near normal conditions −0.99 to 0.99

Results and Discussion

Study Area and Climate Data

This study utilized monthly precipitation data of Sagar and Chattarpur district of Bundelkhand region (Fig. 1) for a 52-year period (1951–2002) acquired from the India Water Portal website. The choice of test stations was ideal for testing the skill of the drought model. In terms of its climatic conditions, the Sagar district had average annual precipitation 1,060 mm while the Chattarpur district had average precipitation 1,032 mm with coefficient of variation 0.211 and 0.257, respectively. Hence, the EDI models developed in this study can be useful tools for drought-risk studies in these sites.

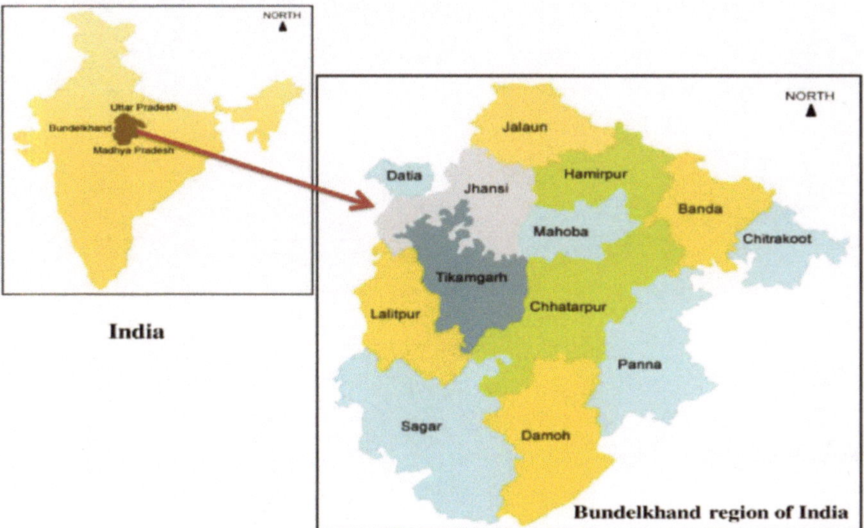

Fig. 1: Geographical location of Bundelkhand region.

Input Selection and Drought Model Development

When designing a robust drought model, it is important to consider pertinent factors such as the proper selection of predictor variables or their combinations, training of predictor variables ensuring the minimum level of over-fitting, selecting the best training algorithm and activation functions ensuring the smallest generalization error, and using the best performance assessment metrics. As there is no "rule-of-thumb" to determine the significant inputs (Tiwari and Adamowski 2013), patterns engrossed in the EDI signal were recognized by correlation statistics which are autocorrelation function (ACF) and partial ACFs. The statistical method attempted to extract lagged information from the signal to analyze time periods between the current index value, and the index at a point in the past domain. Subsequently, optimum inputs for each time lag were identified by statistically analyzing the lagged combinations and correlation coefficients (r). While the ACF detected seasonality in inputs, the PACF removed any dependence on intermediate elements (those within lags), interpreted as a regression of the time-series against its past lagged value. Figures 2 to 5 show correlograms of ACF and PACF for both districts considered with a 95% confidence interval. As a norm, scaling of predictor variables was applied prior to the simulations in order to avoid the patterns or attributes with larger numeric ranges dominating those of the smaller numeric range.

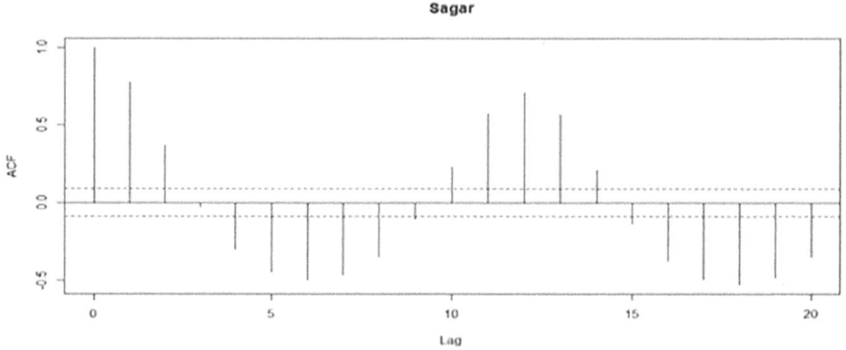

Fig. 2: ACF of Sagar district.

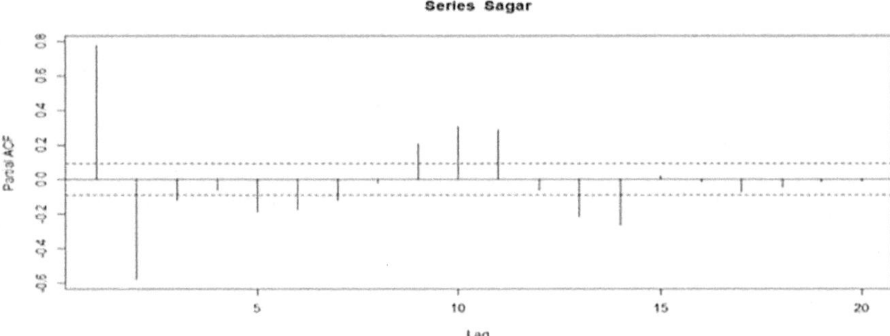

Fig. 3: PACF of Sagar district.

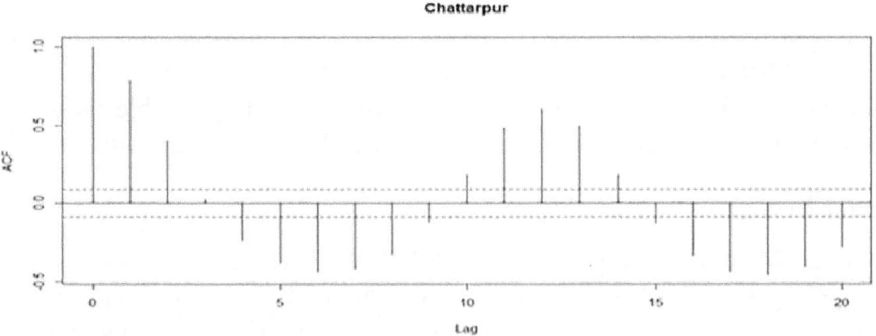

Fig. 4: ACF of Chattarpur district.

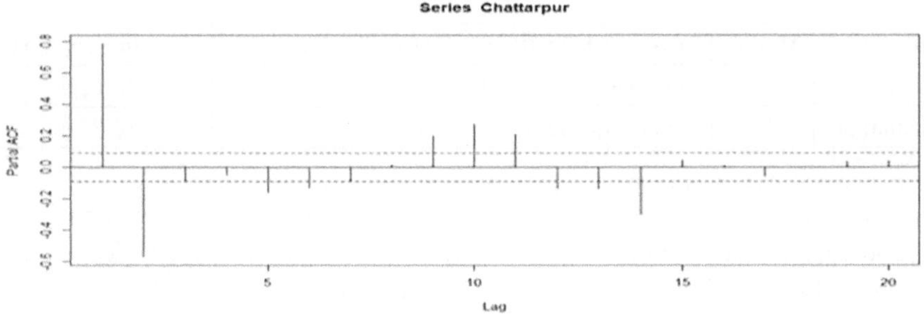

Fig: 5: PACF of Chattarpur district.

The ANN models were developed by considering different architectures with iteratively varying hidden neurons (1–35). During this period, the MSE value was monitored at the iterations of the training phase. The training was stopped when the MSE was at its minimum in comparison to earlier iterations or when the maximum number of epochs was reached (Bishop, 1995).

The LSSVR model developed in this study was based on the radial basis kernel function (RBF). The choice of the RBF was appropriate, as it has a good ability to map nonlinear input samples to high dimensional space, it is able to discover non-linear relationships between target data and input data attributes, and it has superior performance compared to the linear kernel. Also, the linear kernel is a special case of the RBF and the sigmoid kernel behaves like the RBF for certain parameters. The setting of C and r is important for a model's forecasting accuracy. A range of kernel widths, $r = 2^{[15,13,11,9,7,5,3,1,-1,-3,-5]}$ and regularization constants, $C = 2^{[-15,-13,-11,-9,-7,-5,-3,-1,1,3]}$ were trialed to create the best LSSVR structure. The optimized parameters were chosen based on the smallest MSE. Initially, a grid-search was used to identify plausible parameters that were then fine-tuned using the Nelder–Mead simplex algorithm.

Model Evaluation

The available data of 52 years (1951–2002) were divided for two phases, viz the training (1951–1999) and the testing (2000–2002) phases. The training dataset was used for designing both network models. The final outcomes were compared with the EDI's observed (actual) values.

Table 1: The neuronal arrangements for the ANN model.

Location	Model	Training Algorithm or Activation Function	Neuron Structure
	ANN	Levenberg-Marquardt	10-26-1
Chattarpur	ANN	Levenberg-Marquardt	10-32-1

Because it is a well-known universal estimator, the ANN model can rather be considered as standard benchmark. In accordance with Maier and Dandy (2000), all data prior to its inclusion into the ANN model was scaled appropriately. In the present investigation, the input neurons were scaled in the range of [−1, 1] and a transfer function was implemented to explain the nonlinear relationship between input and output neurons. For determining the optimum ANN model to be used in this work, the set of five backpropagation training algorithms used were as follows: scaled conjugate gradient, one-step secant, BFGS quasi-Newton, Bayesian regulation, and Levenberg-Marquardt. Additionally, the set of two commonly used family of hidden transfer functions (hyperbolic-tangent sigmoid, and log-sigmoid) and the three output functions (linear, hyperbolic-tangent sigmoid, and log-sigmoid) were all tried on the testing datasets one at a time in order to seek the optimum model for the final experiments. The number of neurons in the hidden layer was varied gradually but this time over the range 4 to 50 neurons. Consequently, the optimum training algorithm was the Levenberg–Marquardt, the hidden transfer function was the hyperbolic-tangent sigmoid, and the output transfer function was linear with an ANN architecture of 10-26-1 for Sagar and 10-32-1 for Chattarpur district.

The estimation capability of the monthly EDI from the all machine learning algorithms were statistically evaluated using the Root-Mean Square Error (RMSE) and Mean Absolute Error (MAE). Table 2 shows the performance capability of the SVR and ANN models used in this study. The results confirm that ANN model perform better compared to the SVR model in all the evaluation criteria.

Table 2: Performance capability of the SVR and the ANN models.

Location	Predictive Model	RMSE	MAE	Correlation Coefficient
Sagar	ANN	0.690	0.390	0.818
	SVR	0.699	0.435	0.801
Chattarpur	ANN	0.743	0.429	0.761
	SVR	0.778	0.457	0.750

R Code for ANN Model

```
# Load required libraries
library(neuralnet)
library(readxl)
# Load data
data <- read_excel("drought_data.xlsx")
# Split data into training and testing sets
trainIndex <- 1:100
trainData <- data[trainIndex,]
testData <- data[-trainIndex,]
# Define the time series data function
ts.data <- function(data, lag){
 x <- matrix(NA, ncol = lag, nrow = nrow(data)-lag)
 y <- matrix(NA, ncol = 1, nrow = nrow(data)-lag)
 for (i in 1:(nrow(data)-lag)){
 x[i,] <- as.numeric(data[i:(i+lag-1), 2])
 y[i,] <- as.numeric(data[i+lag, 2])
 }
 return(list(x = x, y = y))
}
```

```
# Prepare the training data
ts.train <- ts.data(trainData, lag = 10)
train.x <- ts.train$x
train.y <- ts.train$y
# Define the neural network model
model <- neuralnet(train.y ~ train.x, hidden = c(10, 5), linear.output = TRUE,
 threshold = 0.01, act.fct = "logistic",
 algorithm = "L-BFGS-B", err.fct = "sse",
 startweights = NULL, stepmax = 0.1)
# Prepare the testing data
ts.test <- ts.data(testData, lag = 10)
test.x <- ts.test$x
test.y <- ts.test$y
# Make predictions using the trained model
predictions <- compute(model, test.x)
# Calculate the accuracy of the predictions
accuracy <- sum(round(predictions$net.result) == test.y)/nrow(test.y)
```

References

Bishop C.M. 1995. Neural Networks for Pattern Recognition. Oxford: Oxford University Press.

Byun H.R., Wilhite D.A. "Objective Quantification of Drought Severity And Duration." Journal of Climate 12, (1999): 2747–56

Kim, D.W., Byun, H.R., Choi, K.S. "Evaluation, Modification, and Application of the Effective Drought Index to 200-Year Drought Climatology of Seoul, Korea." Journal of Hydrology 378, (2009): 1–12.

Lippman, R. An introduction to computing with neural nets, IEEE ASSP Magazine 4, (1987): 4–22.

Maier H.R., Dandy, G.C. "Neural Networks for the Prediction and Forecasting of Water Resources Variables: A Review of Modelling Issues and Applications." Environmental Modelling & Software 15, (2000): 101–24.

McCulloch, W.S., Pitts, W. "A Logical Calculus Of The Ideas Immanent In Nervous Activity." The Bulletin of Mathematical Biophysics 5, (1943): 115–33.

Morid, S., Smakhtin, V., Bagherzadeh, K. "Drought Forecasting Using Artificial Neural Networks and Time Series of Drought Indices." International Journal of Climatology 27, (2007): 2103–11.

Tiwari, M.K., Adamowski, J. "Urban Water Demand Forecasting and Uncertainty Assessment Using Ensemble Wavelet-Bootstrap-Neural Network Models." Water Resources Research 49, (2013): 6486–507.

12

Rainfall Prediction Using Neural Networks

Veershetty[1], Harish Nayak G H[1], G. Avinash[1], Vinay H T[2] and Moumita Baishya[1]

[1]*The Graduate School, ICAR-Indian Agricultural Research Institute, Library Avenue, New Delhi 110 012, India*

[2]*Department of Agricultural Statistics, Uttar Banga Krishi Viswavidyalaya Pundibari, Coochbehar 736 165, India*

Abstract

Precise and accurate rainfall estimation is a crucial environmental issue in the present situation. Due to the non-linear and dynamic nature of the rainfall process, complex and effective models and simulations are required for forecasting. Machine learning and deep learning models have demonstrated higher accuracy in forecasting. This study aimed to implement two such models, Generalized Regression Neural Networks (GRNN) and Recurrent Neural Network (RNN), for monthly rainfall data over 115 years in the coastal region of Karnataka. The models were built using training and test datasets, and their forecast performance was compared based on Root Mean Squared Error (RMSE) and Mean Absolute Error (MAE) metrics. Lower values of these metrics indicate better model performance. Among the two fitted models, the RNN model was found to be the best with the lowest RMSE (165.62) and MAE (35.90). Therefore, it can be concluded that the RNN model outperformed the GRNN model for the data used in this study.

Keywords: *Forecasting, Generalized Regression Neural Network, Recurrent Neural Network, and Rainfall*

Introduction

Rainfall is a crucial component of the Earth's climatic system that affects various aspects of human life. It has a direct impact on the ecosystems, water resources, agricultural production, and overall water management. Dependence on rainfall is particularly high in regions where it is the primary source of water, making it essential to have an accurate understanding of future rainfall patterns. Given its importance, rainfall forecasting is a critical concern for a wide range of stakeholders, including governments, businesses, risk management organizations, and the scientific community. However, predicting rainfall is a complex task as it is a random occurrence influenced by numerous factors (Samantaray *et al.*, 2020). Even in similar weather conditions, the possibility of rainfall may vary. This makes historical data and predictive analytics crucial in forecasting future rainfall and taking necessary actions to manage its effects (Avinash *et al.*, 2022). The impact of rainfall extends beyond water management to affect various human activities, including power generation, construction, tourism, and forestry, among others. Therefore, it is crucial to accurately forecast and analyze rainfall patterns to help mitigate the risks and consequences associated with it (Hernandez *et al.*, 2016).

Historically, weather forecasting has been reliant on meteorologists using their experience and knowledge to make predictions based on weather charts and theory (Goswami, 1996). However, the unpredictability of rainfall and its strong association with natural disasters such as flooding, landslides, and avalanches make it essential to develop more accurate forecasting techniques. This has led to the development of numerical weather forecasting using machine learning and deep learning algorithms. The goal of rainfall forecasting is to predict the state of the atmosphere in a given location and numerous approaches have been employed (Nayak *et al.*, 2012), such as machine learning, deep learning, artificial intelligence, data mining, modelling, and statistics. Despite these efforts, predicting rainfall remains a significant challenge due to its nonlinear and unpredictable nature. Accurate rainfall forecasting is crucial to taking preventive and mitigating measures against these natural phenomena. There are two fundamental approaches to predicting rainfall: the dynamical approach and the empirical approach. By implementing suitable approaches for rainfall prediction, society can be better prepared for and respond to natural disasters caused by rainfall.

In order to forecast rainfall, two main approaches are typically employed: the dynamical approach and the empirical approach. The dynamical approach involves using physically built models based on equations of the system to forecast rainfall. This approach is often used by meteorologists to predict weather conditions through mathematical equations that approximate temperature change, pressure change, and other relevant factors. On the other hand, the empirical approach relies on past weather data and their relationship with atmospheric variables across different regions. Regression, machine learning, and deep learning techniques are commonly used in this approach, with neural networks being a key player in prediction. However, choosing the appropriate neural network model depends on the nature of the data being analyzed. Recurrent neural networks are often favored for handling time series data efficiently, making them a suitable option for rainfall forecasting (Prasetya and Djamal, 2019).. In some cases, even unexpected heavy rainfall can be predicted using multilayer perceptron, referred to as "guerrilla rainstorms" in Japan (Kashiwao *et al.*, 2017).

The GRNN and back propagation algorithm are two popular models used in forecasting applications (Wang and Sheng, 2010). GRNN stands out due to its parallel structure, allowing for learning in one iteration and convergence with few training samples (Manek and Singh, 2016). As the data set grows, the error approaches zero, making GRNN a widely used approach in forecasting applications (Nagamulla *et al.*, 2012). In this study, both models were utilized to predict rainfall in the coastal region of Karnataka using 115 years of historical data (1901-2015).

Material and Methods

Recurrent Neural Network (RNN) Model

Recurrent Neural Networks (RNNs) have emerged as a powerful deep learning technique for modeling sequence data. Unlike traditional feed forward neural networks, RNNs have the ability to process sequential data by leveraging their internal memory. This feature enables RNNs to capture dependencies between elements in a sequence, making them particularly useful in applications such as natural language processing, speech recognition, and time series forecasting.

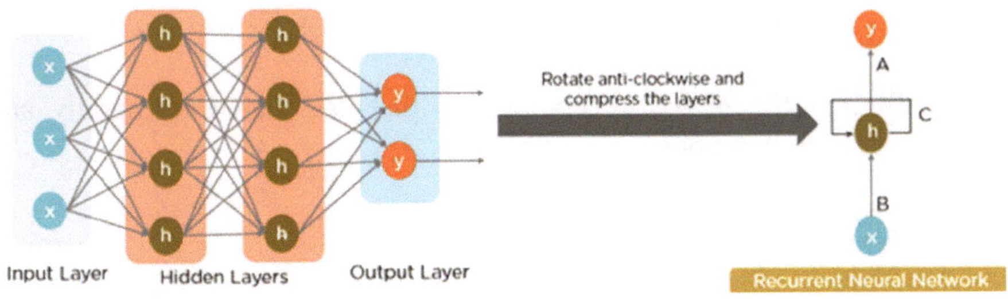

Source: Quora.com

In standard neural networks, each input and output is treated as an independent entity. However, in many real-world scenarios, the previous inputs or outputs are often interdependent, and the past information needs to be remembered to make accurate predictions. This is where RNNs come into play, as they use a hidden layer to store and propagate information across time steps. The hidden state of an RNN encodes a summary of the past inputs and is used to compute the output at each time step.

One of the key advantages of RNNs is that they can handle inputs of varying lengths, making them suitable for modeling sequences of arbitrary length. Additionally, RNNs share the same set of weights across all time steps, allowing them to learn and generalize patterns from sequential data. Overall, RNNs are a powerful class of models that have enabled significant advances in many areas of machine learning and artificial intelligence.

Architecture of a Traditional RNN

RNNs are a type of neural network that has hidden states and allows past outputs to be used as inputs.

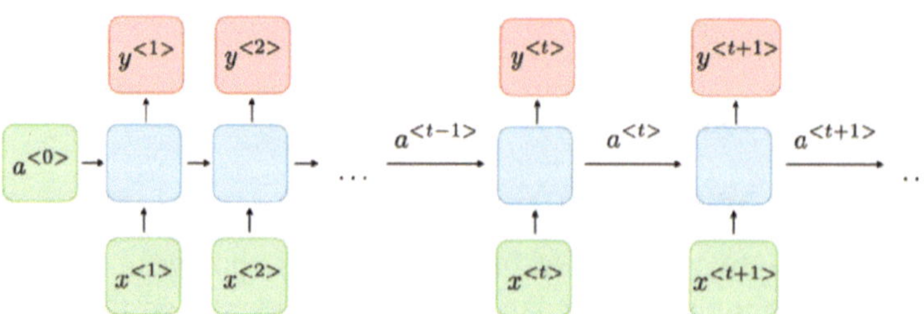

Source: Standford.edu

For each timestep t, the activation $a^{\langle t \rangle}$ and the output $y^{\langle t \rangle}$ are expressed as follows:

$$a^{\langle t \rangle} = g_1\left(W_{aa}a^{\langle t-1 \rangle} + W_{ax}x^{\langle t \rangle} + b_a\right) \text{ and } y^{\langle t \rangle} = g_2\left(W_{ya}a^{\langle t \rangle} + b_y\right)$$

where $W_{ax}, W_{aa}, W_{ya}, b_a, b_y$ are coefficients that are shared temporally and g_1, g_2 activation functions.

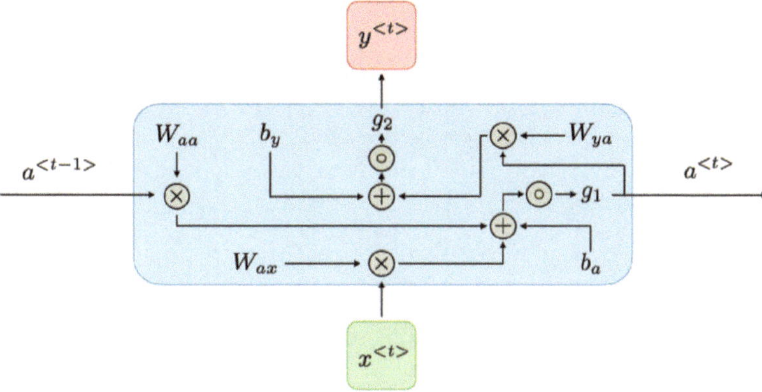

Source: Standford.edu

RNNs are a versatile neural network architecture that can be adapted to various problems depending on the desired input-output relationship. The architecture of RNNs can be modified to suit the specific requirements of the task at hand. Let's take a closer look at some of the commonly used RNN architectures:

One To One: This architecture involves a single input that produces a single output. This is similar to traditional neural networks, which take a fixed input and produce a corresponding output.

One To Many: In this case, a single input can produce multiple outputs. For example, in music generation, a single note can generate an entire melody.

Many To One: Here, multiple inputs from different time steps are combined to produce a single output. Sentiment analysis and emotion detection tasks use this architecture where a sequence of words determines the class label.

Many To Many: This architecture has several variations, where multiple inputs produce multiple outputs. For instance, in machine translation, an English sentence can be translated into a French sentence and vice versa, using many-to-many networks.

By understanding these different RNN architectures, it becomes easier to choose the appropriate model for a particular problem.

Working Process of RNN

The information in recurrent neural networks cycles through a loop to the middle-hidden layer.

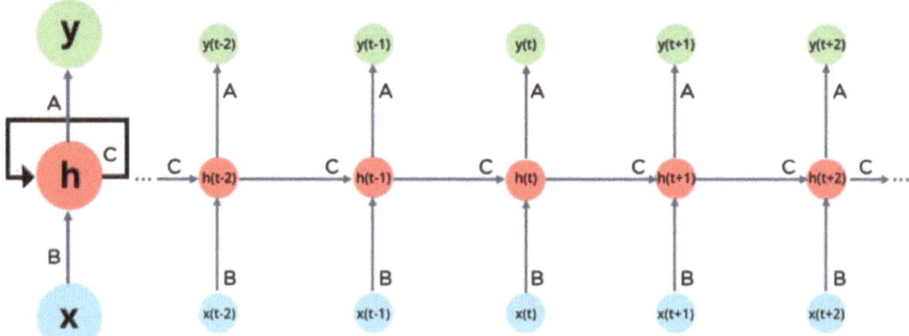

Source: Simplilearn.com

In a neural network, the input layer X is responsible for receiving and processing the input data. The processed data is then passed on to the middle layer h, which can contain multiple hidden layers with unique activation functions, weights, and biases. However, if a neural network lacks memory and the parameters of the hidden layers are not affected by the previous layer, a recurrent neural network is employed.

The Recurrent Neural Network standardizes the different activation functions, weights, and biases of each hidden layer to ensure they have the same characteristics. Rather than creating multiple hidden layers, it creates only one and loops over it as needed. This approach eliminates the need for multiple identical hidden layers, resulting in a more efficient and streamlined architecture.

Common Activation Functions

A neuron's activation function dictates whether it should be turned on or off. Nonlinear functions usually transform a neuron's output to a number between 0 and 1 or -1 and 1.

A Brief Overview of Recurrent Neural Networks (RNN)

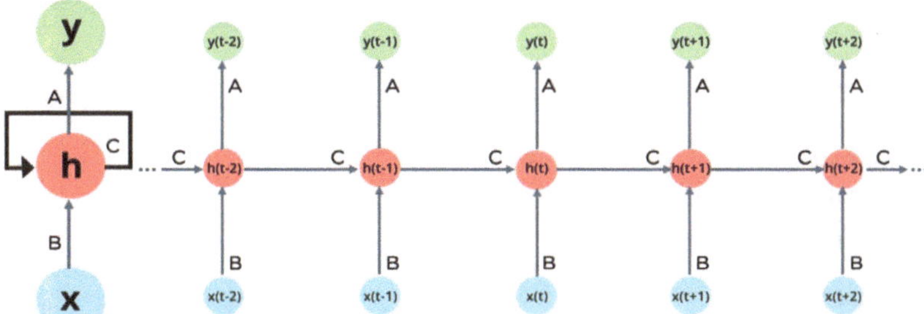

Source: MLTutorial.com

The following are some of the most commonly utilized functions:

- Sigmoid: The formula $g(z) = 1/(1 + e^{\wedge} - z)$ is used to express this.
- Tanh: The formula $g(z) = (e^{\wedge} - z - e^{\wedge} - z)/(e^{\wedge} - z + e^{\wedge} - z)$ is used to express this.
- Relu: The formula $g(z) = \max(0, z)$ is used to express this.

R Codes for RNN Model

```
# Load necessary libraries
library(timeSeries)
library(tseries)
library(aTSA)
library(forecast)
library(rugarch)
library(ModelMetrics)
library(keras)

# Read data from CSV file
data <- read.csv(file.choose(), header = TRUE)

# Split data into training and test sets
```

```
training_data <- ts(data[1:1368, 2])
test_data <- ts(data[1369:1380, 2])

# Plot the data
plot(data[, 2])

# Test for stationarity
adf.test(data[, 2])
pp.test(data[, 2])

# Determine lag order
acf(data[, 2])
pacf(data[, 2])

# Create RNN model
maxlen <- 7
exch_matrix <- matrix(0, nrow = length(training_data) - maxlen - 1, ncol = maxlen + 1)
for(i in 1:(length(training_data) - maxlen - 1)){
  exch_matrix[i,] <- training_data[i:(i+maxlen)]
}
x_train <- exch_matrix[, -ncol(exch_matrix)]
y_train <- exch_matrix[, ncol(exch_matrix)]
x_train <- array_reshape(x_train, dim = c((length(training_data) - maxlen - 1), maxlen, 1))

model <- keras_model_sequential()
model %>%
  layer_dense(input_shape = dim(x_train)[-1], units = maxlen) %>%
  layer_simple_rnn(units = 10) %>%
  layer_dense(units = 1)

# Compile the model
model %>% compile(
  loss = "mse",
  optimizer= "adam",
  metrics = "mae"
)

# Train the model
history <- model %>%
  fit(x_train, y_train, epochs = 100, batch_size = 1, validation_split = 0.1)

# Save the model
save_model_hdf5(model, "rnn_model.h5")

# Load the model
```

```
rnn_model <- load_model_hdf5("rnn_model.h5")

# Make predictions on the test set
maxlen <- 7
exch_matrix2 <- matrix(0, nrow = length(data[,2]) - maxlen - 1, ncol = maxlen + 1)

for(i in 1:(length(data[,2]) - maxlen - 1)) {
  exch_matrix2[i,] <- data[i:(i + maxlen), 2]
}
x_train2 <- exch_matrix2[, -ncol(exch_matrix2)]
y_train2 <- exch_matrix2[, ncol(exch_matrix2)]

x_train2 <- array_reshape(x_train2, dim = c((length(data) - maxlen), maxlen, 1))
pred <- rnn_model %>% predict(x_train2)

# Evaluate the model
df_eval_rnn <- tibble::tibble(y_rnn = y_train2[(length(y_train2) - 11):length(y_train2)],
                   yhat_rnn = as.vector(pred)[(length(y_train2) - 11):length(y_train2)])

# Write the evaluation results to an Excel file
library(writexl)
write_xlsx(df_eval_rnn, "H:\\R codes\\RNNtest1.xlsx")
```

Generalized Regression Neural Network (GRNN) Model

A general regression neural network is a variant of a Radial Basis Function (RBF) network characterized by a fast single-pass learning (Orr, 1996). A GRNN is a machine learning technique which consists of a hidden layer with RBF neurons. Normally, the hidden layer has so many neurons as training examples. The centre of a neuron is its associated training example and so, its output gives a measure of the closeness of the input vector to the training example. Commonly, a neuron will use the multivariate Gaussian function:

$$G(x, x_i) = \exp\left(-\frac{\|x - x_i\|^2}{2\sigma^2}\right)$$

where x_i and σ are the center and the smoothing parameter respectively (x is the input vector).

Given a training set consisting of n training patterns (vectors $\{x_1, x_2, \ldots x_n\}$) and their associated n targets, normally scalars $(\{y_1, y_2, \ldots y_n\})$, the GRNN output for an input pattern x is computed in two steps. First, the hidden layer produces a set of weights representing the closeness of x to the training patterns:

$$w_i = \frac{\exp\left(-\frac{\|x - x_i\|^2}{2\sigma^2}\right)}{\sum_{j=1}^{n} \exp\left(-\frac{\|x - x_j\|^2}{2\sigma^2}\right)}$$

Note that the weights decay with distance to the training pattern. The weights sum to one and represent the contribution of every training pattern to the final result. The GRNN output layer computes the output as follows:

$$\hat{y} = \sum_{i=1}^{n} w_i y_i$$

Therefore a weighted average of the training targets is obtained, where the weights decay with distance to the training patterns. The smoothing parameter controls how many targets are important in the weighted average. When σ is very large the result is close to the mean of the training targets because all of them have a similar weight. When σ is small only the closest training targets to the input vector have significant weights.

Generalized Regression Neural Network

```
# Load necessary libraries
library(timeSeries)
library(tseries)
library(aTSA)
library(forecast)
library(rugarch)
library(ModelMetrics)
library(keras)
library(kernlab)
library(tsfgrnn)

# Read data from CSV file
data <- read.csv(file. Choose(), header = TRUE)

#Combine the stationarity tests and lag order tests into a single function to reduce duplication:
stationarity_test <- function(data) {
  adf <- adf.test(data)
  pp <- pp.test(data)
```

```
  acf <- acf(data)
  pacf <- pacf(data)
  return(list(adf = adf, pp = pp, acf = acf, pacf = pacf))
}

# Call the stationarity_test function
test_results <- stationarity_test(data[, “rainfall”])

# Access individual test results
adf_result <- test_results$adf
pp_result <- test_results$pp
acf_result <- test_results$acf
pacf_result <- test_results$pacf

# Print or inspect the results
print(adf_result)
print(pp_result)
print(acf_result)
print(pacf_result)

##Model devolpment
grnn_model <- grnn_forecasting(training_data, h = 12, lags = 1:12, sigma = 100)
grnn_model$prediction
forecast_data <- predict(grnn_model, h = 12)
autoplot(grnn_model)
```

Illustration

Monthly Rainfall Forecasting for Coastal Karnataka Using Neural Networks: A Data-Driven Approach

Forecasting rainfall is a challenging task due to the high level of nonlinearity in rainfall patterns. Understanding the nonlinear nature of rainfall is crucial for developing accurate forecasting models that can help mitigate the risks associated with low and heavy rainfall. Heavy rainfall can lead to natural disasters like floods that can submerge land used for agriculture and infrastructure, while low rainfall can make it difficult to meet the water needs of people in areas without irrigation. By analyzing the relationships between meteorological factors and rainfall, we can gain insights into the complex interactions that affect rainfall patterns. Sophisticated forecasting models such as Recurrent Neural Network (RNN) and Generalized Regression Neural Network (GRNN) can be used to capture the nonlinear relationships between meteorological factors and rainfall. These models can provide reliable forecasts that aid in planning for and managing the impacts of rainfall.

Studying rainfall is crucial for predicting future scenarios, implementing effective measures to manage low rainfall such as irrigation, and conserving water using water conservation methods. To forecast rainfall accurately in Coastal Karnataka, monthly rainfall data (in mm) was collected from the publicly available database “data.world”. The dataset from 1901 to 2015 and was divided into two parts: the training dataset, which covers the period from 1901 to 2014, and the testing dataset, which covers the year 2015. This division was made to facilitate the development and validation of accurate forecasting models using neural network techniques.

The RNN model’s performance depended on the selection of activation function, number of input nodes, optimizer ‘Adam’ and loss function ‘MSE’, while the GRNN model’s performance was affected by the user-defined hyperparameter ‘sigma’ and the number of lags, which was set to twelve for monthly data. By comparing the performance of these two models, we aimed to provide a comprehensive analysis of their effectiveness in forecasting monthly rainfall in Coastal Karnataka. We used the RNN model for forecasting and implemented it using the ‘TensorFlow’ and ‘Keras’ libraries in R-software. The model was configured with ReLu activation function, 10 input nodes, 3 batch size, 100 epochs, Adam optimizer, and Mean Squared Error (MSE) loss function.

Table 1: Forecast values of different neural networks

Sl. No.	**Actual Rainfall (cm)**	**Predicted Rainfall (cm)**	
		RNN	**GRNN**
1	30.8	46.6	9.9
2	1.4	49.6	12.4
3	1	54.0	1.1
4	32.3	13.3	13.7
5	72.2	137.6	121.4
6	150.3	288.5	654.5
7	735.3	480.9	1163.2
8	930.9	468.0	935.1
9	575.2	480.3	310.8
10	260.3	296.2	144.8
11	208.5	212.8	52.8
12	124.2	226.7	13.9
RMSE		**165.6**	216.0
MAE		**35.9**	115.4

GRNN model: GRNN model was fitted using ***'tsfgrnn'*** package available in R-software.

Lags = 12

The forecast performance of all the fitted models was compared based on RMSE and MAE values on testing dataset.

Result shows that RNN model outperforms better compared to GRNN. The accuracy measures like RMSE and MAE were calculated. Deep learning models are best compare to machine learning due to absence of manual feature extraction.

Conclusion

Among the considered neural network techniques, RNN performs better as compared to GRNN. Further studies can be carried out using other deep learning techniques to model the rainfall.

References

Avinash, G., Ramesh, A. G., Rebasiddanavar, R. M., Shilpa, M., & Vinay, H. T. (2022). A study on temporal variation of land use pattern in Karnataka through indices.

Goswami, B. N. (1996). The challenge of weather prediction. Resonance, 1, 8–17.

Kashiwao, T., Nakayama, K., Ando, S., Ikeda, K., Lee, M., & Bahadori, A. (2017). A neural network-based local rainfall prediction system using meteorological data on the internet: A case study using data from the Japan meteorological agency. Applied Soft Computing, 56, 317-330.

Manek, A. H., & Singh, P. K. (2016). Comparative study of neural network architectures for rainfall prediction. In 2016 IEEE Technological Innovations in ICT for Agriculture and Rural Development (TIAR) (pp. 171-174).

Nayak, D. R., Mahapatra, A., & Mishra, P. (2013). A survey on rainfall prediction using artificial neural network. International Journal of Computer Applications, 72, 16.

Orr, J. L. M. (1996). Introduction to Radial Basis Function Networks. Technical Report, Centre for Cognitive Science, University of Edinburgh, Scotland.

Prasetya, E. P., & Djamal, E. C. (2019). Rainfall forecasting for the natural disasters preparation using recurrent neural networks. In 2019 International Conference on Electrical Engineering and Informatics (ICEEI) (pp. 52-57).

Samantaray, S., Tripathy, O., Sahoo, A., & Ghose, D. K. (2020). Rainfall forecasting through ANN and SVM in Bolangir Watershed, India. In Smart intelligent computing and applications (pp. 767-774).

Wang, Z. L., & Sheng, H. H. (2010). Rainfall prediction using generalized regression neural network: case study Zhengzhou. In 2010 International conference on computational and information sciences (pp. 1265-1268).

Zhou, J., Zhang, Y., Tian, S., & Lai, S. (2020). Forecasting rainfall with recurrent neural network for irrigation equipment. In IOP Conference Series: Earth and Environmental Science (Vol. 510, No. 4, p. 042-040).

13

Multivariate Time Series Analysis for Prediction of Rainfall—A Machine Learning Approach

Harish Nayak G H[1], G Avinash[1], Veershetty[1], Moumita Baishya[1] and Vinay H T[2]

[1]*The Graduate School, ICAR-Indian Agricultural Research Institute, Library Avenue, New Delhi, India – 110 012*

[2]*Department of Agricultural Statistics, Uttar Banga Krishi Viswavidyalaya Pundibari, Coochbehar-736 165*

Abstract

Rainfall estimation is one of the most important and difficult problems in today's world. Because rainfall is highly nonlinear and dynamic, forecasting necessitates sophisticated and powerful models and simulations. Machine learning models have been shown to have higher forecasting accuracy. In this study, we attempted to implement three machine learning models, namely Artificial Neural Networks (ANNs), Support Vector Machine (SVM), and Random Forest (RF), for the monthly rainfall data of Australia over 13 years that are capable of capturing the data's non-linearity. The training dataset was used to build the model, and the testing dataset was used to validate it. The forecast performance of all the selected models was compared based on Root Mean Squared Error (RMSE) and Mean Absolute Percentage Error (MAPE) metrics, lower the value of these metrics better will be the performance of fitted models. In comparison among all the three fitted models, the SVM model was found to be the best fitted model in terms of lowest RMSE (16.72) and MAPE (60.13). Therefore, machine learning models can be used to predict rainfall more precisely.

Keywords: *Artificial Neural Network, Forecasting, Rainfall, Random Forest, Support Vector Machine*

Introduction

Time series data is data that has been collected in a specific pattern with time stamps. (Nayak and Ghosh, 2013). This type of time-oriented data is collected at regular intervals, such as hourly, daily, monthly, or weekly. Time series data can be used to make accurate predictions in a variety of fields and domains, such as foreign currency rates, stock market trends, energy consumption estimates, and climate change. Machine learning and data mining techniques can be used to uncover hidden patterns in historical data and forecast future trends (Aftab *et al.* 2018a, 2018b, Mishra *et al.* 2017).

Weather forecasting based on historical data is a complex but very beneficial task that comes with several issues that must be resolved to achieve optimal results (Wu and Chau, 2013). Weather data includes attributes or features such as temperature, pressure, humidity, and wind speed. Machine learning techniques are used to forecast future weather conditions by identifying hidden patterns and relationships in historical weather data (Aftab *et al.* 2018b). Prediction of precipitation is an important part of the weather forecasting process. The use of machine learning and artificial intelligence techniques is regarded as a critical component in farmer services and policy formulation. Farmers require weather forecasting to plan their activities following the predicted weather. Accurate and timely rainfall prediction in farming, in particular, can be extremely beneficial in planning and security measures for operations, agricultural tasks, water reservoir systems, and construction and transportation activities (Chau and Wu, 2010). Forecasting rainfall will become increasingly important in the future.

Material and Methods

Artificial Neural Network (ANN) Model

ANN algorithms are inspired by the human brain to performs a particular task or functions. ANN perform computations through a process by learning. The neural network is a set of connected input/output units in which each connection has a weight associated with it. In the learning phase, the network learns by adjusting the weights to predict the correct class label of the given inputs.

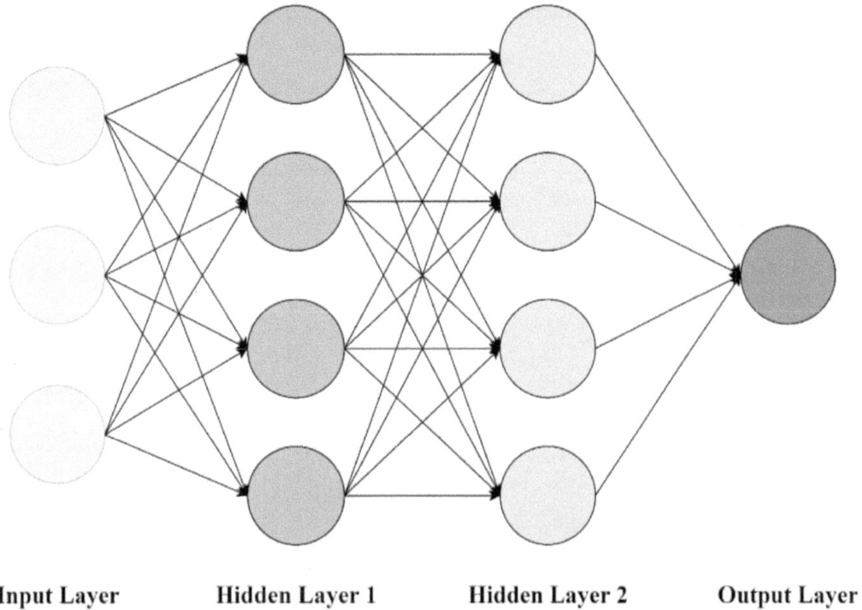

Fig. 1: ANN Architecture

$$Y = \sum(weight * input) + bias$$

Here, $X_{1,} X_{2......} X_n$ are input variables. w_1, w_2.... w_n are weights of respective inputs. The bias, b, is added to the weighted inputs to form the net inputs. Bias and weights are both adjustable parameters of the neuron. Parameters are adjusted using some learning rules. The output of a neuron can range from -inf to +inf. The neuron doesn't know the boundary. So, we need a mapping mechanism between the input and output of the neuron. This mechanism of mapping inputs to output is known as Activation Function.

Activation Functions: Activation function defines the output of a neuron in terms of a local induced field. Activation functions are a single line of code that gives the neural nets non-linearity and expressiveness. There are many activation functions. Some of them are as follows,

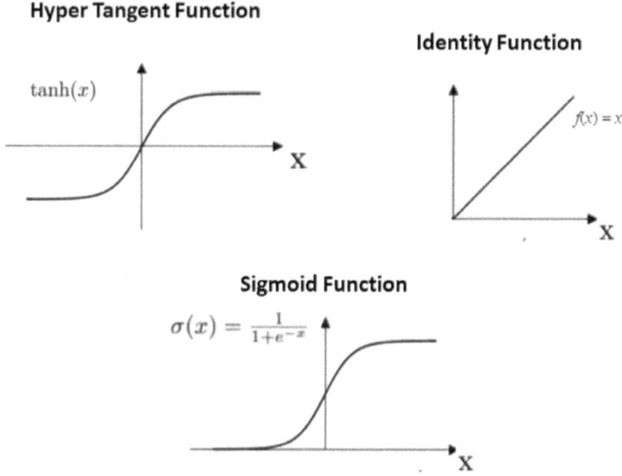

Fig. 2: Activation functions (Source: Datacamp)

```
R Codes for ANN model
# Multivariate Time Series Monthly Rainfall Data
### Artificial Neural Network ###
#Install required packages
install.packages("caret")
install.packages("Metrics")
install.packages("neuralnet")
install.packages("progress")
install.packages("writexl")
# Load required libraries
library(caret)
library(Metrics)
library(neuralnet)
library(progress)
library(writexl)
# Read data
data <- read.csv("C:\\Users\\haris\\OneDrive\\Desktop\\weatherAUS.csv", header = TRUE)
# Split data into training and testing sets
datatrain <- data[1:117, ]
datatest <- data[118:129, ]
# Min-Max normalization process
max_vals <- apply(data, 2, max)
min_vals <- apply(data, 2, min)
# Extract numeric columns (excluding "Date")
numeric_data <- data[, sapply(data, is.numeric)]
# Scale the numeric columns
```

```
scaled <- as.data.frame(scale(numeric_data, center = apply(numeric_data, 2, min), scale =
apply(numeric_data, 2, function(x) max(x) - min(x))))
# Combine the scaled numeric columns with the non-numeric column "Date"
scaled_data <- cbind(data[["Date"]], scaled)
# Prepare training and testing sets after scaling
train.cv <- scaled_data[1:117, ]
test.cv <- scaled_data[118:129, ]
# Cross-validation of neural network model
cv.error <- numeric()
k <- 5  # 5-fold cross-validation
pbar <- progress_bar$new(
  format = "[:bar] :percent :elapsed",
  total = k,
  clear = FALSE,
  width = 60
)
for (i in 1:k) {
  # Min-Max normalization process for each iteration
  max <- apply(numeric_data, 2, max)
  min <- apply(numeric_data, 2, min)
  # Scaling the data
  scaled <- as.data.frame(scale(numeric_data, center = min, scale = max - min))
  train.cv <- scaled[1:117, ]
  test.cv <- scaled[118:129, ]
  # Build neural network model
  nn <- neuralnet(Rainfall ~ ., data = train.cv, hidden = 20, act.fct = "tanh", linear.output =
TRUE)
  # Activation Function: "tanh", "logistic", linear = "f(x){x}"
  # Predict on the training set
  pr.nn <- predict(nn, train.cv)
  pr.nn <- pr.nn * (max(data$Rainfall) - min(data$Rainfall)) + min(data$Rainfall)
     train.cv.r <- (train.cv$Rainfall) * (max(data$Rainfall) - min(data$Rainfall)) +
min(data$Rainfall)
  # Calculate RMSE and store in cv.error
  cv.error[i] <- sqrt(mean((train.cv.r - pr.nn)^2))
  pbar$tick()
}
# Display cross-validation results
cv.error
# Predict on the test set
predict_testNN <- predict(nn, test.cv)
# Unscale the predicted values
predict_testNN <- predict_testNN * (max(data$Rainfall) - min(data$Rainfall)) +
min(data$Rainfall)
# Calculate RMSE and MAPE metrics
rmse <- sqrt(mean((datatest$Rainfall - predict_testNN)^2))
```

```
rmse
mape <- mape(datatest$Rainfall, predict_testNN)
mape
# Export results to Excel
write_xlsx(data.frame(datatest), "H:\\R codes\\ANNactual1.xlsx")
write_xlsx(data.frame(pr.nn), "H:\\R codes\\ANNtrain1.xlsx")
write_xlsx(data.frame(predict_testNN), "H:\\R codes\\ANNtest1.xlsx")
```

Support Vector Machine (SVM) Model

Vapnik and Chervonenkis (1964) proposed a method for developing non-linear classifiers to maximize the margin of the hyperplanes by employing the kernel trick. Cortes and Vapnik proposed the current incarnation (soft margin) in 1993, and it was published in 1995. The mathematical formulation of the Support Vector Regression (SVR) model is explained by Smola and Scholkopf (2004). Suppose the training data given are $\{(x_1, y_1), \ldots, (x_l, y_l)\} \subset \mathcal{X} \times \mathbb{R}$ where x denotes the space of the input pattern. The function f can be described as

$$f(x) = \langle w, x \rangle \text{ with } w \in \mathcal{X}, b \in \mathbb{R}$$

where $<\cdot,\cdot>$ denotes the dot product in $\mathcal{X}$. The small w can be obtained by minimizing the norm (ie) $\| w \| = \langle w, w \rangle$. This problem can be written as a convex optimization problem

$$\text{minimize} \frac{1}{2} \| w \|^2$$

$$\text{subject to} \begin{Bmatrix} y_i - \langle w, x_i \rangle - b \leq \varepsilon \\ \langle w, x_i \rangle + b - y_i \leq \varepsilon \end{Bmatrix}$$

After introducing the slack variables, we get

$$w = \sum_{i=1}^{l} \left(\alpha_i - \alpha_i^* \right) x_i$$

And the function f becomes

$$f(x) = \sum_{i=1}^{l} \left(\alpha_i - \alpha_i^* \right) \langle x_i, x \rangle + b.$$

The Karush-Kuhn-Tucker condition was used to compute b. Following the computation of b and the construction of the regression model, the examples with non-vanishing coefficients are referred to as support vectors. A greater number of support vectors accurately explains the relationships. The kernel concept is used by the support vector method to convert the given data into a higher dimension. Kernels are classified into four types: linear, polynomial, sigmoid, and radial basis function kernels. The radial basis function kernel is used for numerical analysis.

$$K(x_i, x_{i'}) = \exp\left(-\gamma \sum_{j=1}^{p} (x_{ij} - x_{i'j})^2\right)$$

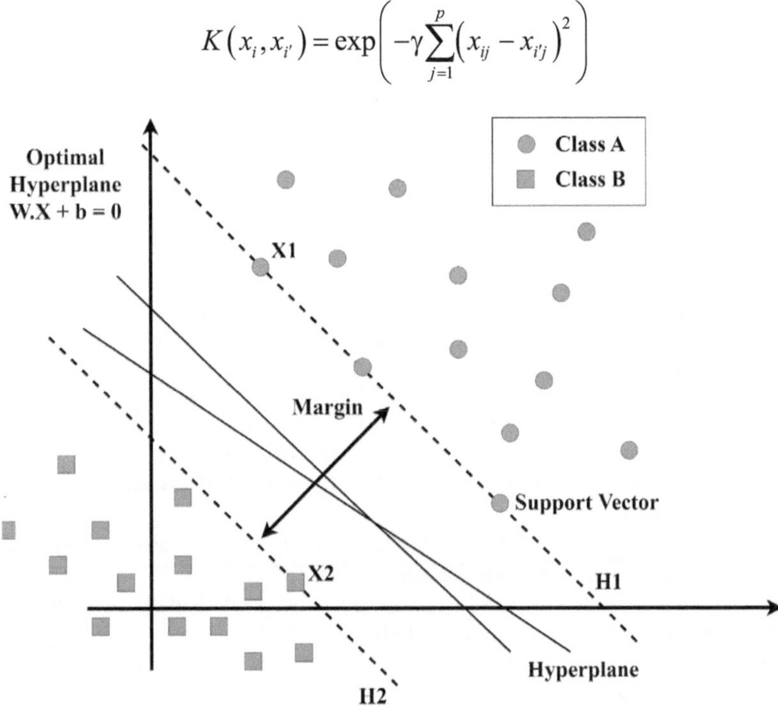

Fig. 3: SVM Architecture

```
### Support Vector Machine ###
# Load required libraries
library(e1071)
library(writexl)
# Read data
data <- read.csv("C:\\Users\\haris\\OneDrive\\Desktop\\weatherAUS.csv", header = TRUE)
# Create training and test set
datatrain <- data[1:117, ]
datatest <- data[118:129, ]
# Extract numeric columns (excluding "Date")
numeric_data <- data[, sapply(data, is.numeric)]
```

```
# Min-Max normalization process
max_vals <- apply(numeric_data, 2, max)
min_vals <- apply(numeric_data, 2, min)
scaled <- as.data.frame(scale(numeric_data, center = min_vals, scale = max_vals - min_
vals))
trainNN <- scaled[1:117, ]
testNN <- scaled[118:129 , ]
# Tuning the hyperparameters C, gamma, epsilon through Grid Search optimization
# 5 fold cross-validation technique
# It takes time to run depending upon the size of the data (more time if data is large) and
#number of hyperparameters
# kernels – radial, linear, sigmoid, polynomial
# Install and load the e1071 package
install.packages("e1071")
library(e1071)
# Tuning the model
tuneResult <- tune(svm, Rainfall ~ ., kernel = "radial", data = trainNN,
            ranges = list(epsilon = seq(0, 1, 0.1), cost = 1:100, gamma = seq(0, 5, 0.1)),
            tunecontrol = tune.control(cross = 5))
tunedModel <- tuneResult$best.model
print(tunedModel)
tunedModelY <- predict(tunedModel, trainNN) # Train data-Fitted values
tunedModelY1 <- predict(tunedModel, testNN) # Test data-Predicted values
# Unscaling the data
predict_trainNN <- (tunedModelY * (max(numeric_data$Rainfall) - min(numeric_
data$Rainfall))) + min(numeric_data$Rainfall)
predict_testNN <- (tunedModelY1 * (max(numeric_data$Rainfall) - min(numeric_
data$Rainfall))) + min(numeric_data$Rainfall)
# Calculation of RMSE and MAPE metrics
rmse <- sqrt(mean((datatest$Rainfall - predict_testNN)^2))
rmse
mape <- mape(datatest$Rainfall, predict_testNN)
mape
# Export results to Excel
install.packages("writexl")
library(writexl)
# Export actual test data
write_xlsx(data.frame(datatest), "H:\\R codes\\SVRactual1.xlsx")
# Export fitted values
write_xlsx(data.frame(predict_trainNN), "H:\\R codes\\SVRtrain1.xlsx")
# Export predicted values
write_xlsx(data.frame(predict_testNN), "H:\\R codes\\SVRtest1.xlsx")
```

Random Forest (RF) model

A Random Forest, as the name implies, is a tree-based ensemble in which each tree is determined by a set of random variables. More formally, for a p-dimensional random vector $X = (X_1, \ldots, X_p)^T$ representing the real-valued input or predictor variables and a random variable Y representing the real-valued response, we assume an unknown joint distribution $P_{XY}(X, Y)$. The goal is to find a prediction function $f(X)$ for predicting Y (Cutler *et al.* 2011). The prediction function is determined by a loss function $L(Y, f(X))$ and defined to minimize the expected value of the loss

$$E_{XY}\left(L(Y, f(X))\right)$$

where the subscripts denote expectation with respect to the joint distribution of X and Y.

Intuitively, $L(Y, f(X))$ is a measure of how close $f(X)$ is to Y; it penalizes values of $f(X)$ that are a long way from Y. Typical choices of L are squared error loss $L(Y, f(X)) = (Y - f(X))^2$ for regression and zero-one loss for classification:

$$L(Y, f(X)) = I(Y \neq f(X)) = \begin{cases} 0 \text{ if } Y = f(X) \\ 1 \text{ otherwise.} \end{cases}$$

It turns out (see, for example, [10] Sect. 2.4) that minimizing $E_{XY}\left(L(Y, f(X))\right)$ for squared error loss gives the conditional expectation

$$f(x) = E(Y \mid X = x)$$

otherwise known as the regression function. In the classification situation, if the set of possible values of Y is denoted by $\mathcal{Y}$, minimizing $E_{XY}\left(L(Y, f(X))\right)$ for zero-one loss gives

$$f(x) = \arg\max_{y \in \mathcal{Y}} P(Y = y \mid X = x),$$

otherwise known as the Bayes rule.

Ensembles construct f in terms of a collection of so-called "base learners" $h_1(x), \ldots, h_J(x)$ and these base learners are combined to give the "ensemble predictor" $f(x)$. In regression, the base learners are averaged

$$f(x) = \frac{1}{J} \sum_{j=1}^{J} h_j(x),$$

while in classification, $f(x)$ is the most frequently predicted class ("voting")

$$f(x) = \underset{y \in \mathcal{Y}}{\text{argm}} \sum_{j=1}^{J} I\left(y = h_j(x)\right).$$

In Random Forests, the th base learner is a tree denoted $h_j(X, \Theta_j)$, where Θ_j is a collection of random variables and the Θ_j's are independent for $j = 1,\ldots,J$. Although the definition of a Random Forest is very general, they are almost invariably implemented in the specific way (Cutler *et al.* 2011). To understand the Random Forest algorithm, it is important to have a fundamental knowledge of the type of trees used as base learners.

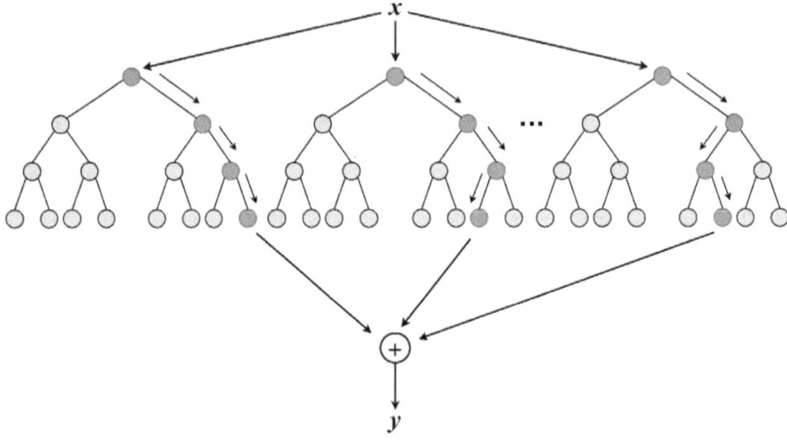

Fig. 4: RF Architecture

```
### Random Forest ###
# Load required libraries
install.packages("randomForest")
library(randomForest)
library(writexl)
# Read data
data <- read.csv("C:\\Users\\haris\\OneDrive\\Desktop\\weatherAUS.csv", header = TRUE)
# Create training and testing set
datatrain <- data[1:117, ]
datatest <- data[118:129, ]
```

```
# Extract numeric columns (excluding "Date")
numeric_data <- data[, sapply(data, is.numeric)]
# Min-Max normalization process
max_vals <- apply(numeric_data, 2, max)
min_vals <- apply(numeric_data, 2, min)
scaled <- as.data.frame(scale(numeric_data, center = min_vals, scale = max_vals - min_
vals))
trainNN <- scaled[index, ]
testNN <- scaled[-index, ]
# Define repeated cross-validation with 5 folds and three repeats
repeat_cv <- trainControl(method = 'repeatedcv', number = 5, repeats = 3)
# 'mtry' is the hyperparameter, which is in the hands of the investigator
# 'ntree' can also be changed in the sequence (100,200,300,400,500)
randomForestModel <- randomForest(Rainfall ~ ., data = trainNN, ntree = 400, mtry = 3,
importance = TRUE, na.action = na.omit, trControl = repeat_cv, metric = 'mean_squared_
error')
# Fitted values
predict_testNN1 <- predict(randomForestModel, trainNN)
# Unscaling the data
predict_testNN1 <- (predict_testNN1 * (max(numeric_data$Rainfall) - min(numeric_
data$Rainfall))) + min(numeric_data$Rainfall)
# Predicted values
predict_testNN2 <- predict(randomForestModel, testNN)
# Unscaling
predict_testNN2 <- (predict_testNN2 * (max(numeric_data$Rainfall) - min(numeric_
data$Rainfall))) + min(numeric_data$Rainfall)
# Calculation of RMSE and MAPE metrics
rmse <- sqrt(mean((datatest$Rainfall - predict_testNN2)^2))
rmse
mape <- mape(datatest$Rainfall, predict_testNN2)
mape
# Export the actual test data into Excel
install.packages("writexl")
library(writexl)
model <- data.frame(datatest)
write_xlsx(model, "H:\\R codes\\RFactual1.xlsx")
# Export fitted values
model <- data.frame(predict_testNN1)
write_xlsx(model, "H:\\R codes\\RFtrain1.xlsx")
# Export predicted values
model <- data.frame(predict_testNN2)
write_xlsx(model, "H:\\R codes\\RFtest1.xlsx")
```

Illustration

Forecasting of Monthly Rainfall of Australia Using Machine Learning Models

Broadly, Australia can be divided into two seasonal rainfall regimes: North and the South. The Northern half of the continent typically experiences a monsoonal, summer wet season, with rainfall falling from October through April, and the rest of the year being dry. The southern half of the continent—particularly southern parts of Western Australia and South Australia, and the whole of Victoria and Tasmania—experiences its highest rainfall during the cooler months from April through November. Rainfall in Australia is highly variable, largely because of large-scale atmospheric and oceanic drivers that affect the region. The most important driver of Australian rainfall variability is ENSO, particularly for rainfall across eastern Australia, excluding Tasmania. This leads to very high year-to-year rainfall variability, such that long-term trends because of climate change are harder to distinguish than for temperature. Therefore, it necessitates to forecast the rainfall before and accurately. For the present study, we have considered the monthly rainfall data of Australia for the period of 13 years *i.e.*, from 2008 to 2019 consisting of different variables *viz.*, Rainfall, Minimum Temperature, Maximum Temperature, Evaporation, Wind Gust Speed, Sunshine, Wind Speed at 9am, Wind Speed at 3pm, Humidity at 9am, Humidity at 3pm, Pressure at 9am, Pressure at 3pm, Temperature at 9am, Temperature at 3pm among which we have considered Rainfall as a dependent variable and all others as independent variables and have tried to fit the competitive models namely Artificial Neural Networks (ANNs), Support Vector Machine (SVM), and Random Forest (RF).

In our study, ANN, SVM and RF were fit to monthly rainfall data of Australia. The training dataset was used for building the model and testing dataset was used for validating the forecast performance of the fitted model. It is known that ANN model depends on activation function (hyperbolic, logistic, linear) and the number of hidden nodes (1 to 20), but there is no specific rule to choose them while SVM depends on the Kernel function (Radial Basis Function (RBF), sigmoid, polynomial, linear) and hyperparameters C [1, 100], gamma [0, 10], epsilon [0, 1] which are user defined, however Grid Search optimization technique allows to tune the hyperparameters and obtain the optimal hyperparameters. RF model depends on hyperparameter 'mtry' which is user defined and 'ntree' can also be changed accordingly.

The optimal hyper-parameters used for forecasting using different models:

All the following models are fitted using 5-fold cross validation technique to find the optimal combinations of the parameters.

ANN model: ANN consist of one input layer with 13 input nodes, one hidden layer with 18 hidden nodes, one output layer with one output node. For building ANN model, *Resilient Backpropagation* neural network, which is available in ***'neuralnet'*** package of R-software was used.

Activation function = Hyperbolic (tanh), Number of hidden nodes = 18

Table 1: Forecast values of different machine learning models

Sl. No.	Actual Rainfall (cm)	Predicted Rainfall (cm)		
		ANN	SVR	RF
1	50.80	67.82	65.36	42.19
2	48.37	21.15	44.16	14.60
3	18.83	45.89	17.02	76.52
4	15.63	21.62	28.66	35.82
5	9.67	9.87	5.52	66.77
6	22.45	9.08	12.23	17.45
7	44.64	32.33	26.68	55.67
8	4.86	8.35	19.72	90.32
9	94.37	51.05	53.53	30.25
10	53.55	93.23	63.25	64.48
11	65.21	51.46	47.66	28.22
12	21.58	23.33	36.33	42.61
RMSE		22.66	**16.72**	42.52
MAPE		65.17	**60.13**	263.42

SVM model: SVM model was fitted using ***'e1071'*** available in R-software.

Kernel = RBF, $C = 2$, gamma = 0.2, epsilon = 0.1

RF model: RF model was fitted using ***'randomForest'*** available in R-software.

mtry = 3, ntree = 400

The forecast performance of all the fitted models was compared based on RMSE and MAPE values on testing dataset.

Based on the results from the above table we can say that, the SVM model is performing better *i.e.,* the actual and predicted values are nearer compared to other fitted models. Through RMSE and MAPE values, the superiority of the SVM model can be depicted.

Conclusion

Among the considered machine learning techniques, SVM performs better as compared to other techniques. The accuracies of other techniques such as ANN and RF up to the mark.

References

A. J. Smola, and B. Scholkopf, "A tutorial on support vector regression", Statistics and Computing, vol.14, pp. 199-222, 2004 ·

Aftab, S.; Ahmad, M.; Hameed, N.; Salman, M.; Ali, I.; Nawaz, Z. Rainfall Prediction in Lahore City using Data Mining Techniques. Int. J. Adv. Comput. Sci. Appl. 2018a, 9, 254–260.

Aftab, S.; Ahmad, M.; Hameed, N.; Salman, M.; Ali, I.; Nawaz, Z. Rainfall Prediction using Data Mining Techniques: A Systematic Literature Review. Int. J. Adv. Comput. Sci. Appl. 2018b, 9, 143–150.

Chau, K.W.; Wu, C.L. A hybrid model coupled with singular spectrum analysis for daily rainfall prediction. J. Hydroinformatics, 2010, 12, 458–473.

Cutler, Adele & Cutler, David & Stevens, John. (2011). Random Forests. 10.1007/978-1-4419-9326-7_5.

Gupta, D.; Ghose, U. A comparative study of classification algorithms for forecasting rainfall. In Proceedings of the 2015 4th International Conference on Reliability, Infocom Technologies and Optimization (ICRITO) Trends and Future Directions, Noida, India, 2–4 September 2015; pp. 1–6.

Mishra, N.; Soni, H.K.; Sharma, S.; Upadhyay, A. A Comprehensive Survey of Data Mining Techniques on Time Series Data for Rainfall Prediction. J. ICT Res. Appl. 2017, 11, 168.

Nayak, M.A.; Ghosh, S. Prediction of extreme rainfall event using weather pattern recognition and support vector machine classifier. Arch. Meteorol. Geophys. Bioclimatol. Ser. B 2013, 114, 583–603.

V. J. Yohai, "High breakdown point and high efficiency robust estimates for regression", Annals of Statistics, vol. 15, pp. 642-656, 1987.

V. N. Vapnik, "The Nature of Statistical Learning Theory", Springer, New York, 1995.

V. N. Vapnik, and A. Y. Chervonenkis, "A note on one class of perceptrons", Automation and Remote Control, vol.25, pp. 821-837, 1964

Wu, C.L.; Chau, K.W. Prediction of Rainfall Time Series Using Modular Soft Computing Methods. Eng. Appl. Artif. Intell. 2013, 26, 997–1007.

Wu, J.; Long, J.; Liu, M. Evolving RBF neural networks for rainfall prediction using hybrid particle swarm optimization and genetic algorithm. Neurocomputing, 2015, 148, 136–142.

Yue, T.; Zhang, S.; Zhang, J.; Zhang, B.; Li, R. Variation of representative rainfall time series length for rainwater harvesting modelling in different climatic zones. J. Environ. Manag. 2020, 269, 110731.

14

Drought Coping Mechanisms: An Investigation of Determinants of Adoption by Decision Learning Approach

Halagundegowda G R[1], Singh A[2], Naveena K[3] and Nagaraja M S[4]

[1]Central Silk Board, Bengaluru

[2]Banaras Hindu University, Varanasi

[3]Centre for Water Resources Development and Management, Kunnamangalam Kozhikode

[4]Keladi Shivappa Nayaka University of Agricultural and Horticultural Sciences, Shivamogga

---

Abstract

Drought can happen when a region receives below-average precipitation, resulting in prolonged shortages in its water supply. It can have a substantial impact on the ecosystem of the affected region. When drought occurs in a particular area, it obviously affects crop and livestock production. To reduce the effect of drought on farm production and to stabilize farm income, farmers have to take some systematic measures, and such measures are called drought coping mechanisms. The present study aims to identify the determinants of drought coping mechanisms adopted by the farmers to mitigate the impact of drought on their farming activity in Kolar district of Karnataka. The decision tree (CHAID) was carried out by considering the various socio-economic characteristics of farmers as predictors and the adoption behavior of the farmers as the response variable. The result of the decision tree shows that the variables such as crop insurance, extension visits, crop diversification, and Income were significant and they are influencing variables in the adoption of drought coping strategies. The model showed 96.7% of classification prediction.

Keywords: *CHILD, decision tree, drought coping, farmer*

Introduction

Drought is defined as "when a region receives below average precipitation, resulting in prolonged shortages in its water supply, whether surface or ground water or atmospheric. It can have a substantial impact on the ecosystem and agriculture of the affected region." When drought arises in a particular area, obviously it affects the crop and livestock production. In order to reduce the effect of drought on farm productivity and to stabilize farm income, farmers have to take some systematic measures, and such measures are referred to as drought coping mechanisms. The objective of the study was to identify the determinants of the adoption of drought-coping mechanisms to balance and stabilize the farm income of the stakeholders. The study also helps to know how to mitigate the effect of drought on farmer's livelihoods.

The existing literature on the adoption of drought coping mechanisms is quite meager. Akwango *et al.*, (2016), Chandan Kumar Jha *et al.*, (2017), Emmanuella Ellis (2017), Ronald Twongyirwe *et al.*, (2019) assessed the adoption strategies by considering multinomial logistic regression and the results state that the use of different tree planting, crop varieties, soil and water conservation, early and late planting, and furrow irrigation. The data related to the adoption of any agriculture technology have qualitative response variables with two or more categories. Most of the studies used qualitative response models such as the logit model, probit model, and multivariate techniques like discriminant analysis to measure the farmer's perception toward the adoption of any agriculture technology. In this study, we chose machine learning models such as Decision tree, to assess the classification ability and to determine the factors influencing adoption.

The outcome of the study would be useful to the policymakers to implement drought relief programs during the drought periods depending upon the intensity of drought and its impact on levels of income and productivity. Identification of factors determining the adoption of drought coping strategies would be helpful to focus more on the factors to speed up the adoption rate. Estimates of the economics of production would be helpful to breeders and extension personnel to demonstrate the benefits of growing drought coping strategies.

Materials and Methods

Decision Tree

The decision tree is a decision support tool that uses a tree-like graph or model of decisions and their possible consequences, including chance event outcomes, resource costs, and utility. It is one way to display an algorithm. Decision trees are commonly used in operations research, specifically in decision analysis, to help identify a strategy most likely to reach a goal.

The decision tree is a flowchart-like structure in which each internal node represents a "test" on an attribute (e.g., whether a coin flip comes up heads or tails), each branch represents the outcome of the test and each leaf node represents a class label (decision taken after computing all attributes). The paths from root to leaf represent classification rules. In decision analysis, a decision tree and the closely related influence diagram are used as a visual and analytical decision support tool, where the expected values (or expected utility) of competing alternatives are calculated. The following aspects of Decision tree are important to be discussed.

(i) Decision rules

The decision tree can be linearized into decision rules, where the outcome is the contents of the leaf node and the conditions along the path form a conjunction in the "if clause." Decision rules can also be generated by constructing association rules with the target variable on the right. In general, the rules have the form: If condition1 and condition 2 and condition 3 then outcome.

(ii) Influence diagram

Much of the information in a decision tree can be represented more compactly as an influence diagram, focusing attention on the issues and relationships between events. The squares represent decisions, the ovals represent action, and the diamond represents results (Fig. 1).

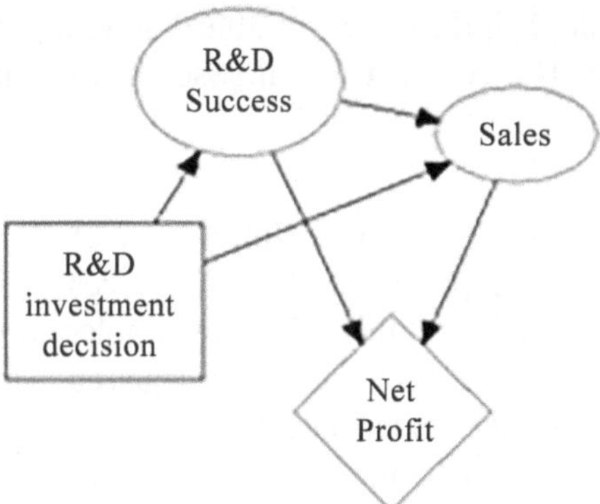

Fig.1: Influence diagram.

(iii) Decision tree learning

Decision tree learning uses a decision tree as a predictive model which maps observations about an item to conclusions about the item's target value. It is one of the predictive modeling approaches used in statistics, data mining, and machine learning. Tree models in which the target variable can take a finite set of values are called classification trees. In these tree structures, leaves represent class labels and branches represent conjunctions of features that lead to those class labels. Decision trees where the target variable can take continuous values (typically real numbers) are called regression trees.

A tree can be "learned" by splitting the source set into subsets based on an attribute value test. This process is repeated on each derived subset in a recursive manner called recursive partitioning. The recursion is completed when the subset at a node has all the same value of the target variable, or when splitting no longer adds value to the predictions. This process of top-down induction of decision trees (TDIDT) is an example of a greedy algorithm, and it is by far the most common strategy for learning decision trees from data. In data mining, decision trees can be described also as the combination of mathematical and computational techniques to aid the description, categorization, and generalization of a given set of data. The data comes in records of the form:

$$(X, Y) = (x_1, x_2, x_3 \dots\dots\dots x_k, Y) \quad (1)$$

The dependent variable, Y, is the target variable that we are trying to understand, classify or generalize. The vector X is composed of the input variables, x_1, x_2, x_3 etc., that are used for that task.

1. **CHAID:** Chi-squared Automatic Interaction Detector Performs multi-level splits when computing classification trees. A CHAID method of growing tree selects the number of input variables corresponding nominal dependent variable. At each step, Chi-squared Automatic Interaction Detection chooses the independent (predictor) variable that has the strongest interaction with the dependent variable. Categories of each predictor are merged if they are not significantly different with respect to the dependent variable.

R-Code for Decision Tree Using Child Method

a. **Import the data**
```
k1<-read.table(file="clipboard",header = TRUE)
```
b. **Create train/test set**
```
data_train <- create_train_test(k1, 0.8, train = TRUE)
data_test <- create_train_test(k1, 0.8, train = FALSE)
install.packages("rpart")
library(rpart)
install.packages("rpart.plot")
library(rpart.plot)
```
c. **Build the model**
```
rpart(formula, data=, method ='')

arguments:
- formula: The function to predict
- data: Specifies the data frame- method:
- "class" for a classification tree
- "anova" for a regression tree

fit <- rpart(survived~., data = data_train, method = 'class')
```
d. **Visualization**
```
tiff("E:\\r codes\\asis sir\\FINAL PCA Results\\dec1.tiff",compression = "lzw", height =
3000, width = 3000,  res = 400)
rpart.plot(fit,extra = 106,box.palette="RdBu", cex = 1.2)
dev.off()
```

Illustration

The Kolar district of south Karnataka which belongs to Eastern Dry Zone and most of farmers were involved in rainfed agriculture and are suffering now because of shortage of rainfall and frequent drought. Hence adoption of certain coping strategies against drought is the major solution to stabilize the farm income during the drought period. The specific reason for choosing this study was to identify the factors influencing on adoption of any strategies against drought and its impact on agriculture policy on Karnataka's agricultural cropping pattern.

Nature and Source of Data

The household data collected on the socio-economic characters of farmers of Kolar districts of Karnataka (India) by employing the multi-stage sampling design, in which the household of a farmer acts as the last unit of the sampling design. The data is mainly related to coping strategies implemented against drought by the farmers of this region. The total sample size (farmers) is 150 and to develop the model, it is necessary to split the set of data into two different subsets. Therefore, the training data consists of 80% of respondents

(120 farmers) and the test data consists of the remaining 20% of respondents (30 farmers). Table 1 gives the variable of interest and the unit of measurement used in this study.

Table 1: Variables encoding summary

Code	Variables	Measurement
Y	Adoption behavior	$Y = 0$ for Non-Adopters $= 1$ for Adopters
X_1	Age of the farmer	Number of years
X_2	Education of the farmer	Formal Years of Education
X_3	Household Size	Number of family members
X_4	Farm Size	Number/quantity of acres
X_5	Farming Experience	Number of years
X_6	Animal Husbandry	Number of farm animals
X^7	Media Exposure	Number of sources exposed frequently
X_8	Extension Visits	Number of Visits
X_9	Crop Diversification	Number of Crops Grown in that year
X_{10}	Income Status	In Rupees (Rs.)
X_{11}	Worth of Liquidating Assets	In Rupees (Rs.)
X_{12}	Crop Insurance by Government	In Rupees (Rs.)

A CHAID method of growing tree was fitted to the data and the number of input variables about 12 to the corresponding nominal dependent variable (Adoption). At each step, Chi-squared Automatic Interaction Detection chooses the independent (predictor) variable that has the strongest interaction with the dependent variable. Categories of each predictor are merged if they are not significantly different with respect to the dependent variable.

Table 2: Model summary

Specifications	Growing Method	CHAID
	<TB>Dependent Variable	Adoption
	Independent Variables	Age, Education, Household Size, Farm Size, Farming Experience, Animal Husbandry, Media Exposure, Extension Visits, Crop Diversification, Income Status, Liquid Assets, Crop Insurance
	Validation	Split Sample
	Maximum Tree Depth	3
	Minimum Cases in Parent Node	100
	Minimum Cases in Child Node	50

Results	Independent Variables Included	Crop Insurance, Ext Visits, Crop Diversification, Income
	Number of Nodes	10
	Number of Terminal Nodes	6
	Depth	3

Table 2 provides broad information about the specifications used to build the model. The specifications sector provides information on the settings used to generate the tree model, including the variables used in the analysis. The results section displays information on the number of total and terminal nodes, the depth of the tree (number of levels below the root node), and independent variables included in the final model. Validation of samples was done by split sample method and the maximum tree depth by default for CHAID method is 3, minimum cases in parent node and minimum cases in child node for CHAID by default are 100 and 50, respectively.

The summary table also indicates that the tree model itself is a particularly simple one since it has 10 nodes and 6 terminal nodes. This may not be an issue if you want a reliable model that can be applied in a practical fashion rather than a simple model that is easy to describe or explain. For practical purposes, a model shouldn't rely on too many independent (predictor) variables. In this case, that's not a problem because 12 independent variables were specified, but only 4 were included in the final model.

The variables such as crop insurance, extension visits, crop diversification, and income had significant contributions to the model; the remaining variables were automatically dropped from the final model. In the other literature on the adoption of drought coping strategies, Francis Ndamani *et al*., (2016), Stephen Abugri *et al*., (2020), and Mequannt Marie *et al*., (2020) used logistic regression and showed that education, household size, annual household income, access to information, credit, and membership of farmer-based organization are the most important factors that influence farmers' adaptation to drought coping strategies.

Figure 2 represents the tree diagram for the training data set, which is a graphical representation of the tree model. Using the CHAID method, crop insurance is the best predictor of the adoption of drought coping mechanisms (adoption behavior of farmers). For the low crop insurance category, crop insurance level is the only significant predictor of the adoption behavior of farmers. Of the farmers in this category, 100% of the farmers were nonadopters. Since there are no child nodes below it, this is considered a terminal node. For the medium crop insurance category (0 to 16,200), 42.9% of farmers who had income ranges up to Rs. 16,200 were nonadopters and 57.1% of farmers in this group were adopters.

In the case of the high crop insurance category (>16,200), 4.2% of farmers having income >Rs.16,200 were nonadopters and 95.8% of farmers were adopters in this group, For the medium crop insurance category, the next best predictor is the number of extension visits, the variable, number of extension visits were splits into the farmers with lower number of visits (<3) and higher number of visits (>3), the farmers with lower number of visits have no further predictor and will wind up as terminal node, whereas the farmers with higher number of visits have further best predictor such as crop diversification (low diversified area if the number of crops is equal to 1, otherwise high diversified area if >1 crops). The high crop insurance category having the next best predictor is income. The income variable is split into a low-income group (< 24,500) and a high-income group (>24,500).

The tree has three splits, each splits acts as separate rule and the rule which runs from Crop Insurance (Rs 16,200) to Extension visits (>3 visits) toward Crop Diversification (>1 Crops) was the best rule because this rule predicts about 87.5% farmers as adopters. Meenakshi *et al.*, (2015) used probit analysis to examine the factors influencing on adoption of drought-tolerant rice varieties, the result shows that farming experience, level of education, access to extension service, and price of output positively and significantly influence the adoption of drought-tolerant rice varieties.

Figure 2 represents the tree diagram for training data set, which is a graphical representation of the tree model. Using the CHAID method, crop insurance is the best predictor of adoption of drought coping mechanisms (adoption behavior of farmers). For the low crop insurance category, crop insurance level is the only significant predictor of adoption behavior of farmers. Of the farmers in this category, 100% of the farmers were nonadopters. Since there are no child nodes below it, this is considered a terminal node. For the medium crop insurance category (0 to 16,200), 42.9% of farmers who had income ranges up to Rs. 16,200 were nonadopters and 57.1% of farmers in this group were adopters.

In the case of the high crop insurance category (>16,200), 4.2% of farmers having income >Rs.16,200 were nonadopters and 95.8% of farmers were adopters in this group. For the medium crop insurance category, the next best predictor is the number of extension visits, the variable, number of extension visits were splits into the farmers with lower number of visits (<3) and higher number of visits (>3), the farmers with lower number of visits have no further

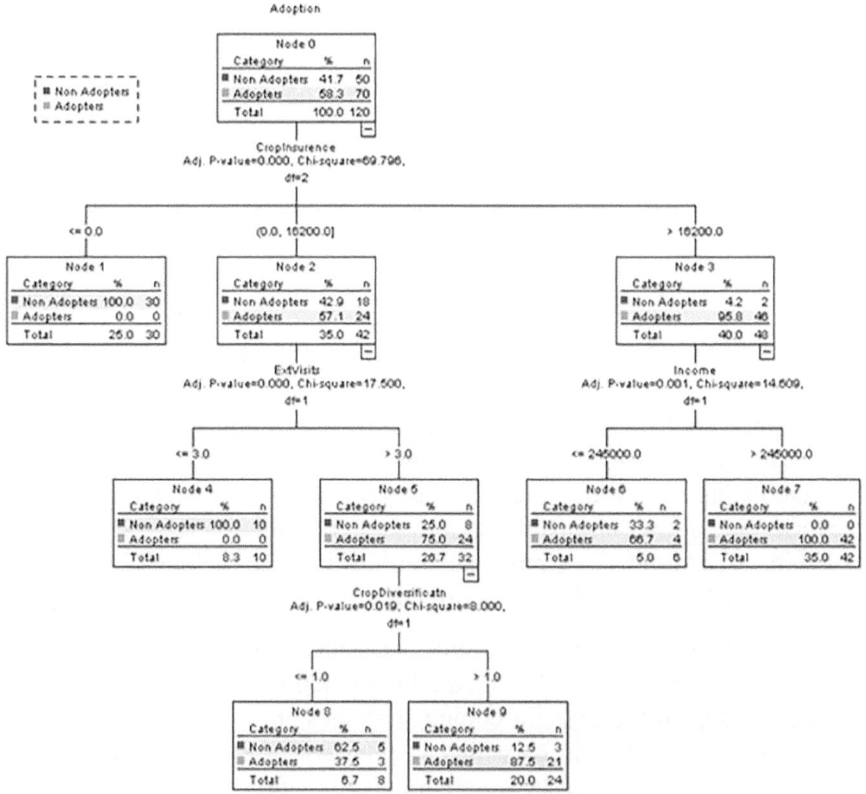

Fig. 2: CHAID Tree chart.

predictor and will wind up as terminal node, whereas the farmers with higher number of visits have further best predictor such as crop diversification (low diversified area if the number of crops is equal to 1, otherwise high diversified area if >1 crop). The high crop insurance category having the next best predictor is income. The income variable splits into a low-income group (<24,500) and high-income group (>24,500).

The tree has three splits, each splits acts as a separate rule and the rule which runs from Crop Insurance (Rs 16,200) to Extension visits (>3 visits) toward Crop Diversification (>1 Crops) was the best rule because this rule predicts about 87.5% farmers as adopters. Meenakshi *et al.*, (2015) used probit analysis to examine the factors influencing the adoption of drought-tolerant rice varieties and the result shows that farming experience, level of education, access to extension service, and price of output positively and significantly influence the adoption of drought tolerant rice varieties.

Table 3: Classification matrix

Sample	Observed	Predicted		
		Nonadopters	**Adopters**	**Percent Correct**
Training	Nonadopters	45	5	90.0%
	Adopters	3	67	95.7%
	Overall Percentage			93.3%
Test	Nonadopters	12	0	100.0%
	Adopters	1	17	94.4%
	Overall Percentage			96.7%

Table 3 provides the classification table for both training data set and testing data set; the classification table helps to assess the performance of the model by cross tabulating the observed response categories with the predicted response categories and shows how well our full model correctly classifies the cases. A classification result is a simple summary of the number and percentage of subjects classified correctly and incorrectly.

The cells off the diagonal of the cross-classification of cases are incorrect predictions of the cases used to create the model, 67 of the 70 farmers who previously adopted the drought coping strategies are classified correctly and only 3 cases were wrongly classified. Forty-five of the 50 nonadopters are classified correctly and 5 cases were wrongly classified. Overall, 93.3% of the training cases are classified correctly, corresponding to the 6.7% incorrect as shown in the model summary table. A better model should correctly identify a higher percentage of the cases. The testing sample helps to validate the model; here 96.7% of these cases were correctly classified by the model. This suggests that the overall current model is correct and efficient in prediction and classification.

Conclusion

The main goal of this research is to find the determinants of adoption of drought coping strategies by the machine learning approach such as decision tree method. Further, the variables such as extension visits, crop insurance, crop diversification, and Income were significant factors influencing on adoption of drought coping strategies by the farmers. The classification tables provide a quick evaluation of how well the model works and 93.3% of the training cases are classified correctly which shows our model performs better. The testing sample supports to validation of the model; here 96.7% of these cases were correctly classified by the model and it shows the selected CHAID model is more than enough to investigate the drivers of the adoption.

References

Abugri, S.A. "A Study on Farmers Perception and Preference in Subscribing to Drought-Index Crop Insurance in the Northern Region of Ghana." Regional Economic Development Research 1, no. 2 (2020):63–76.

Akwango, D.A., Obaa, B.B., Turyahabwe, N., Baguma, Y., Egeru A. "Agro-Pastoral Choice of Coping Strategies and Response to Drought in the Semi-Arid Areas of Uganda." African Journal of Rural Development 1, no. 3 (2016): 281–91.

Ashwinirani, Vidyavathi, B.M. "Ameliorated Methodology for the Design of Sugarcane Yield Prediction Using Decision Tree. COMPUSOFT." An International Journal Of Advanced Computer Technology 4, no. 7 (2015): 1882–89.

Chandan Kumar Jha, Vijaya Gupta, Utpal Chattopadhyay, Binil kumar Amarayil Sreeraman (2017): Migration As Adaptation Strategy to Cope with Climate Changes: A Study Of Farmers' Migration in Rural India. International Journal of Climate Change Strategies and Management 10, (2017): 121–41.

Chauhan, S., Singh, D., Singh, A. "An Improved Feature Selection and Classification Using Decision Tree for Crop Datasets." International Journal of Computer Applications.142, no. 13 (2016): 21–26.

Chen, G., Ma, L. 2008. Research on Rough Set and Decision Tree Method Application in Evaluation of Soil Fertility Level. College of Information and Technology Science, Jilin Agricultural University, Chang Chun Jilin China.

Emmanuella E. "Factors Affecting Risk Management Strategies to Climate Change Effects in Ghana." International Journal of Food and Agricultural Economics 5, no. 1 (2017): 1–17.

Francis Ndamani1and Tsunemi Watanabe. "Determinants of Farmers' Adaptation to Climate Change: A Micro Level Analysis in Ghana." Scientia Agricola 73 no. 3 (2016): 201–8.

Halagundegowda G. R., Singh, A., Meenakshi H. K. "Discriminant Analysis for Classification of Farmers Based on Adoption of Drought Coping Mechanisms." Agriculture Update 12, (2017): 2643–8.

Hanamaratti, N.G., Prashanthi, S.K., Salimath, P.M., Hanchinal, R.R., Mohankumar, H.D., Parameshwarappa, K.G., Raikar, S.D. "Traditional Land Races of Rice in Karnataka: Reservoir of valuable Traits." Current Science, 94, no. 2 (2008): 242–7.

Joshi, K.K., Patidar, P. "An Approach to Construct Decision Tree Using SLIQ and KNN for Land Grading System." International Journal of Engineering, Science and Management 5, no. 1 (2015): 231–6.

Joshi, K.K Patidar, P. "Indian Agriculture Land through Decision Tree in Data Mining." International Journal of Core Engineering & Management 1, no. 5 (2014): 93–8.

Meenakshi, H.K., Halagundegowda G.R., Nagaraja, M. S. "Discriminant Analysis on Adoption of Drought Tolerant Rice Varieties in Kolar District of Karnataka." Trends in Biosciences 8, no. 7 (2015): 1693–6.

Meenakshi, H.K., Selvaraj, K.N., Halagundegowda, G.R., Nagaraja, M.S. "Probit Analysis on Adoption of Drought Tolerant Rice Varieties and Factors Influencing on Its Adoption in Kolar District of Karnataka." Progressive Research-An International Journal, 10, (2015): 2799–801.

Mequannt Marie, Fikadu Yirga, Mebrehtu Haile, Filmon Tquabo. "Farmers Choices and Factor Affecting Adoption of Climate Change Adaptation Strategies: Evidence from Northwestern Ethiopia." Heliyon 6, (2020): 1–10.

Mohammed Nasir Uddin, Wolfgang Bokelmann, Jason Scott Entsminger. "Factors Affecting Farmers' Adaptation Strategies to Environmental Degradation and Climate Change Effects: A Farm Level Study in Bangladesh." Climate 2 (2014): 223–41.

Mulyono, S., Sadly, M., Ivan Fanany, M., Basaruddin, T. "A New Heuristic Decision Tree and Time invariant Detection of Paddy Growth Stages from Modis Data." Journal of Theoretical and Applied Information Technology 71, no. 3 (2015): 429–34.

Nagaratna Biradhar, Sridhar. "A Study on Consequences of 2003 Drought in Karnataka with Reference to Livestock and Fodder." Journal of Human Ecology 26, no. 2 (2004): 123–30.

Ravindra, M., Lokesha, V., Kumara, P., Ranjan, A. "Study and Analysis of Decision Tree Based Irrigation Methods in Agriculture System." International Journal of Emerging Technology and Advanced Engineering 2, no. 12 (2012): 167–73.

Ronald T., David M., Bernard B., Barbara R. N., Hannington O., Viola N., Grace M. "Perceived Effects of Drought On Household Food Security in South-western Uganda: Coping Responses and Determinants." Weather and Climate Extremes 24, (2019):100201.

Roth, H and Botha, N. "Using Ethnographic Decision Tree Modelling to Explore Farmers' Decision-making Processes: A Case Study." New Zealand Agricultural and Resource Economics Society, (2009): 27–8.

Sadhu, S.K., Ramasubramanian, V., Rai, A., Kumar, A. "Decision Tree Based Models for Classification in Agricultural Ergonomics." Statistics and Applications. 12, no. 1 (2014): 21–33.

Sharma, R., Ghosh, A., Joshi, P.K. "Decision Tree Approach for Classification of Remotely Sensed Satellite Data Using Open Source support." Journal of Earth System Science 122, no. 5 (2015): 1237–47.

Veenadhari, S., Mishra, B., Sing, C.D. "Soybean Productivity Modeling Using Decision Tree Algorithms." International Journal of Computer Applications 27, no. 1 (2011): 11–16.

Yi-Yang, G., Nan-Ping, R. "Data Mining And Analysis Of Our Agriculture Based on the Decision Tree." International Colloquium on Computing, Communication, Control, and Management 5, no. 2 (2009): 136–40.

15

Classification of Farmers Based on Drought Coping Strategies Using Support Vector Machine (SVM)

Halagundegowda G R[1], Singh A[2], Naveena K[3] and Nagaraja M S[4]

[1]Central Silk Board, Bengaluru

[2]Banaras Hindu University, Varanasi

[3]Centre for Water Resources Development and Management, Kunnamangalam, Kozhikode

[4]Keladi Shivappa Nayaka University of Agricultural and Horticultural Sciences, Shivamogga

Abstract

The proper classification of farmers based on their adoption of drought coping mechanisms is critical in mitigating the negative impacts of drought on farm income. To achieve this, a study was conducted in the Kolar district of Karnataka to investigate the classification ability of different Support Vector Machine (SVM) architectures, including internal parameters and kernel types. The study found that the SVM with sigmoid kernel trick, with hyperparameter gamma = 0.096 and classification type 1 with capacity C = 2.40, performed well in accurately classifying farmers based on their drought coping strategies. The optimal value of the hyperparameter was computed by running 1,000 iterations for tuning the model using the random grid search optimization approach with sigmoid kernel trick and coefficient value of 0.06. This SVM architecture had 97 overall support vectors, with 93 bounded support vectors. The classification summary showed that 93.3% of the cases were correctly classified, with only 7% being wrongly classified. On the other hand, the SVM with radial basis function (RBF) kernel trick, with hyperparameter gamma = 0.083 and classification type 1 with capacity C = 1.20, was also used in the study. The optimal value of the hyperparameter was computed using the random grid search optimization approach with the RBF kernel trick. This SVM architecture had 82 overall support vectors, with 74 bounded support vectors. However, its classification performance was lower than that of the SVM with sigmoid kernel trick. The findings suggest that the SVM with sigmoid kernel trick is a more suitable model, as it provides better classification accuracy than the SVM with RBF kernel trick for classifying farmers based on their drought coping strategies.

Keywords: *Drought coping mechanisms, Farmers' classification, Kernel types, Kolar district, Random grid search optimization, Support Vector Machines (SVMs)*

Introduction

Drought occurs when a region experiences below average precipitation, resulting in prolonged shortages in its water supply, which can have a significant impact on agriculture and the ecosystem. To reduce the effect of drought on farm production and stabilize farm income, farmers must take systematic measures called drought coping mechanisms. These measures can be categorized into two broad categories: Ex-ante coping mechanisms, which are taken before the occurrence of drought, and Ex-post coping mechanisms, which are taken after the occurrence of drought. Ex-ante coping mechanisms include activities such as crop, spatial, and varietal diversification, income diversification, and the use of savings. Ex-post coping mechanisms include changing consumption patterns, borrowing money, liquidating assets, seeking government drought relief funds, and migration. Drought can have a significant impact on crop and livestock production, causing farmers to adopt drought coping mechanisms to stabilize their farm income. This study goals to identify the factors that determine the adoption of such coping mechanisms, to mitigate the effects of drought on the livelihoods of farmers.

Aryal *et al*., (2019) and Halagundegowda *et al*., (2017) conducted studies to assess the factors affecting the adoption of drought coping mechanisms. Aryal *et al* used a multivariate probit model to assess the probability and level of adoption of multiple climate smart agriculture practices, while Halagundegowda *et al* used discriminant analysis to examine the factors influencing the adoption of drought coping strategies. The results of Halagundegowda *et al*'s study revealed factors such as farm size, extension visits, crop diversification, and crop insurance positively influenced the adoption of drought coping mechanisms, while age had a negative influence. When analyzing the adoption of agricultural technologies, researchers often encounter data with a qualitative response variable containing two or more categories. To determine farmers' perception toward the adoption of such technologies, previous studies have employed qualitative response models such as logit model, probit model, and multivariate techniques such as discriminant analysis. However, in this study, we opted for a machine learning model, specifically Support Vector Machine (SVM), to assess its classification ability in predicting the adoption of agriculture technologies. SVM is a powerful and flexible machine learning algorithm that has shown promising results in various fields, including agriculture.

Materials and Methods

Support Vector Machine for Classification

SVMs are supervised learning models with associated learning algorithms that analyze data used for classification and regression analysis. Given a set of training examples, each marked as belonging to one or the other of two categories, an SVM training algorithm builds a model that assigns new examples to one category or the other, making it a non-probabilistic binary linear classifier.

(i) Linear Support Vector Machine for Separable Data

Vapnik and Chervonenkis (1964) proposed original linear SVM formulation for separable data. For a given binary classification problem, the objective is to estimate functions f with parameter vector θ such that $\{f(x;\theta): R^n \rightarrow \{-1, +1\}\}$ using a finite set of training data. Let the training data set consists of $\{x_i, y_i\}_{i=1}^{N}$ with input patterns $x_i \in R_n$ and their respective class labels $y \in \{-1\ +1\}$. When the training data is linearly separable, a separating hyperplane (a hyperplane that separates the positive from the negative examples, this hyperplane is shown in Fig. 1 of the form

$$w^T x + b = 0 \tag{1}$$

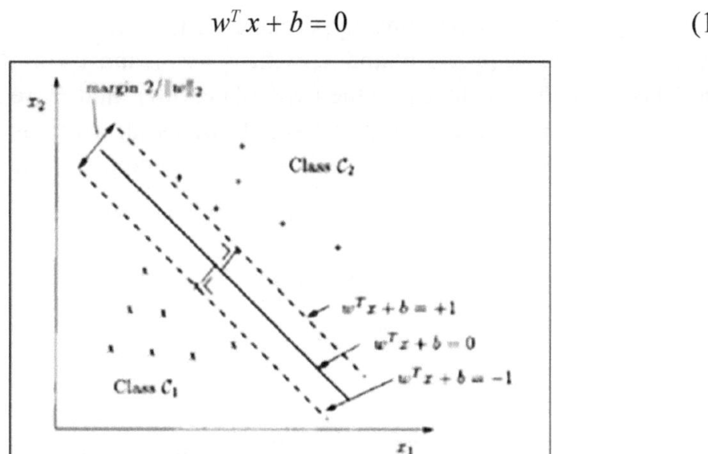

Fig.1: Construction of optimal hyperplane for binary classification problem illustrated for two-dimensional input space.

can be fitted to correctly classify training patterns, where weight vector w is normal to the hyperplane and defines its orientation, b is the bias term, and T is the transpose. From equation (1) a linear classifier (decision function) given by

$$y(x) = \text{sign}(w^T x + b), \tag{2}$$

which classify class C_2 ($y_i = +1$ if $w^T x + b \geq 0$) if and class C_1 ($y_i = -1$ if $w^T x + b \leq 0$) patterns. Let H_+ ($w^T x + b = +1$) and $H\_$ ($w^T x + b = -1$) the shortest distance from the separating hyperplane to the closest class 2 (class 1), then the margin of the hyperplane is defined as the sum of H_+, i.e., $H\_ + H\_$ An optimal hyperplane for linearly separable set of training data is defined as the linear decision function with maximal margin between the vectors of two classes, as is shown Figure 1. The support vector algorithm will construct this optimal separating hyper plane. It is shown that the optimal hyper plane will have good generalization abilities, and only a relatively small amount of training data is needed to construct this plane. The set of margin-determining training vectors are called the support vectors.

Assume all training data satisfy

$$\begin{cases} w^T x_i + b \geq +1 & \text{if } y_i = +1 \\ W^T x_i + b \leq -1 & \text{if } y_i = -1 \end{cases} \quad i = 1,2, \ldots N \tag{3}$$

This can be combined into a single set of equalities:

$$y_i (w^T x_i + b) - 1 \geq 0, \quad i = 1,2,\ldots N, \tag{4}$$

where N is the training data size,

To find the optimal separating hyperplane, it is necessary to maximize the margin $H_+ + H_-$. Suppose x_1 and x_2 with $y_i = +1$ and $y_i = -1$ are Class C_2 and Class C_1 patterns closest to the hyperplane of H_+ and $H\_$ respectively. For maximal separation, the hyperplane should be as far away as possible from each of them. By letting $\|\cdot\|$ be the l_2 norm of a vector, we can get

$$w^T x_1 + b = +1 \text{ and } w^T x_2 + b = -1$$

$$W^T (x_1 - x_2) = +2$$

$$\frac{w}{\|w\|} \cdot (x_1 - x_2) = \frac{2}{\|w\|}$$

where $\|w\| = \sqrt{w^T w} = \sqrt{W_1^2 + W_2^2 + \ldots + W_n^2}$ is the l_2 norm of a vector.

Maximizing the margin is equivalent to maximizing $2/||w||$, which is in turn the same as solving

$$J_P \min_{w,b} \frac{1}{2} ||w||^2 \text{ or } J_P \min_{w,b} \frac{1}{2} w^T w \tag{5}$$

subject to constraints

$$y_i (w^T x_i + b) \geq 1, \quad i = 1,2,\ldots N \tag{6}$$

Constructing the optimal hyperplane is therefore convex Quadratic Programming (QP) problem. We call this optimization problem is in the primal formulation. Therefore SVM formulations are done within a context of convex QP optimization problem. The general methodology is to start formulating the problem in the primal weight space as a constrained optimization problem. Then Lagrange multipliers and Kursh-Kühn-Tucker (KKT) complimentary conditions are used to find the optimal solution. Under condition for optimality, above QP problem is finally solved in the dual space of Lagrange function.

The Lagrangian multipliers $\alpha_i \geq 0$, are introduced for each of the constrains in equation (6) to get the following Lagrangian:

$$L(w,b;\ a) = \frac{1}{2} || w ||^2 - \sum_{i=1}^{N} a_i (y_i (w^T x_i + b) - 1). \tag{7}$$

The objective is to minimize equation (7) with respect to w and b under the requirement that the derivatives of the Lagrangian with respect to all the a_i vanish. This must be subject to the constrains that the Lagrangian multipliers remains non-negative. Since all the constrains are linear in convex quadratic optimizations, it is possible to equivalently solve the dual optimization problem of maximizing equation (7), such that the gradient of L with respect to w and b vanishes, and it is required that $\alpha_i \geq 0$ That is

$$\frac{\partial}{\partial b} L(w,b;\ a) = 0 \text{ and } \frac{\partial}{\partial w} L(w,b;\ a) = 0 \tag{8}$$

and thus

$$\sum_{i=1}^{N} a_i y_i = 0 \text{ and } w = \sum_{i=1}^{N} a_i y_i x_i \tag{9}$$

By substituting equation (8) in (9), the dual form of the optimization problem is derived, which is also QP determined by

$$\mathrm{J}_D \max_a = \sum_i^N a_i - \frac{1}{2} \sum_i^N \sum_j^N a_i a_j y_i y_j x_i^T x_j \tag{10}$$

subject to

$$a_i \geq 0 \text{ and } \sum_{i=1}^{N} a_i y_i = 0; \; i = 1,2,\ldots N. \tag{11}$$

Thus by solving the dual QP problem, the coefficients α_i are obtained. These coefficients are then used to calculate w from equation (11). The vector w will be a solution to problem (11). The decision function from equation (11) can be rewritten as

$$f(x) = sign\left(\sum_{i=1}^{N} a_i y_i x^T x_i + b\right) \tag{12}$$

The decision surface in equation (12) is determined by the N Lagrangian multipliers α_i These multipliers are either zero or positive. The subset of zero multipliers will have no effect on the decision function, and can be omitted. It is the set of positive multiplier that influences the classification, and their corresponding training vectors are called the *support vectors*. The last unknown parameter b is determined by taking average of $b = y_i - \sum_{i=1}^{N} a_i y_i x^T x_i$ for all support vectors.

(ii) Feature Function and Kernel Functions

The idea of transforming the input space into a feature space of a higher dimension by using feature functions $\varphi(x_i)$ and then performing a linear classification in that higher dimensional space is central to SVM. However, the feature space may have a very high dimensionality, even infinite. An obvious consequence is to avoid the inner product of feature functions $\varphi(x_i)$. Fortunately, a method was developed to generate a mapping into a high-dimensional feature space with kernels. The rationale that prompted the use of kernel functions is to enable computations to be performed in the original input space rather the high-dimensional (even infinite) feature space. Using this approach, the SVM algorithm avoids the evaluation of the inner product of the feature functions.

Under certain conditions, an inner product in feature space has an equivalent kernel in input space. For any symmetric continuous function $K(x, z)$ satisfying Mercer's condition, there exists a Hilbert space H, a map $\varphi(x)$: $R^n \rightarrow R^{nH}$ and a positive λ_i number such that one can write $K(x,z) = \sum_{i=1}^{nH} \lambda_i \varphi(x)\varphi(z)$, where $x, z \in R^n$ and n_H is the dimension of H (which can be infinite dimensional). Mercer's condition requires that

$$\int K(x,z) g(x) g(z) dxdz \geq 0 \tag{13}$$

Then *Kernel function* can be expressed as the inner product (often called the dot product) which is given as $K(x,z) = \varphi(x)^T \varphi(z)$, where $\varphi(x) = \sqrt{\lambda_i}\varphi_i(x)$ and $\varphi(z) = \sqrt{\lambda_i}\varphi_i(z)$ The application of kernel function ((*x*, *z*)) is often called the *kernel trick* (Cristianini and Shawe-Taylor, 2000) (Table 1).

Table 1: Some typical choices of kernel function are (Cristianini and Shawe-Taylor, 2000)

Kernel Type	Expression
Linear SVM	$K(x,x_i) = x_i^T x$
Polynomial of degree *d*	$K(x,x_i) = \left(x_i^T x + k\right)^d$
Radial basis function (RBF)	$K(x,x_i) = \exp\ \left\{-\lVert x - x_i \rVert^2 / 2\sigma^2\right\}$
	or Equivalently
Multi-Layer Perceptron (MLP)	$K(x,x_i) = \exp\ \left\{-y \lVert x - x_i \rVert^2\right\}$
	$K(x,x_i) = \tanh\left(k_1 x_i^T x + k_2\right)$

The Mercer condition holds for all σ or γ values in the RBF kernel case and *k* values in the polynomial case but not for all possible choices of, k_1, k_2 in the MLP case.

Table 2: Classification ability measures

		Predicted Class	
Actual Class		Class = Yes	Class = No
	Class = Yes	True Positive (TP)	False Negative (FN)
	Class = No	False Positive (FP)	True Negative (TN)

To evaluate how well our model is performing, we use metrics such as true positive, true negative, false positive, and false negative as given in Table 2. We want to maximize true positives and true negatives while minimizing false positives and false negatives. Based on these values, we can calculate other metrics such as accuracy, precision, recall, and F1 score.

Accuracy: Accuracy is the most intuitive performance measure and it is simply a ratio of correctly predicted observation to the total observations. One may think that, if we have high accuracy then our model is best. Yes, accuracy is a great measure but only when you have symmetric datasets where values of false positive and false negatives are almost same.

$$\text{Accuracy} = TP+TN/TP+FP+FN+TN$$

Precision: Precision is the ratio of correctly predicted positive observations to the total predicted positive observations. The question that this metric answer is of all passengers that labeled as survived, how many actually survived? High precision relates to the low false positive rate.

$$\text{Precision} = TP/TP+FP$$

Recall (Sensitivity): Recall is the ratio of correctly predicted positive observations to the all observations in actual class—yes. The question recall answers are: of all the passengers that truly survived, how many did we label?

$$\text{Recall} = TP/TP+FN$$

F1 score: F1 Score is the weighted average of Precision and Recall. Therefore, this score takes both false positives and false negatives into account. Intuitively it is not as easy to understand as accuracy, but F1 is usually more useful than accuracy, especially if you have an uneven class distribution. Accuracy works best if false positives and false negatives have similar cost. If the cost of false positives and false negatives are very different, it's better to look at both Precision and Recall.

$$\text{F1 Score} = 2*[(\text{Recall} * \text{Precision}) / (\text{Recall} + \text{Precision})]$$

Illustration

This study was conducted in Kolar district, which is situated in the Eastern Dry Zone of Karnataka. Due to the shortage of rainfall and being a drought-prone area, most of the farmers in this area practice rainfed agriculture. Hence, the adoption of coping strategies against drought is crucial for stabilizing farm income during periods of drought. The study aimed to identify the factors that influence the adoption of drought coping strategies and to understand the impact of agricultural policies on cropping patterns in Karnataka. The study also aimed to examine exactly how cropping patterns fluctuate from period to period and area to area during droughts.

Nature and Source of Data

The current study utilized both classification and prediction techniques, with household data used to fit classificatory statistical models. The data was collected on the socio-characteristics of farmers in Kolar district, Karnataka, India, and related to coping strategies implemented against drought. The data was collected using a multi-stage sampling design, which involved selecting Kolar district as the first stage, major rainfed crops growing taluks as the second stage, and major rainfed crops growing villages as the third stage. The data collection process allowed for a representative sample of the population to be selected for analysis.

Table 3: The number of respondents and splitting ratio of training and test data set

Class	Total Respondents	Training set (80%)	Test set (20%)
Non-Adopters	62	50	12
Adopters	88	70	18
Total	150	120	30

The number of respondents in training and test data set are reported in Table 3. In the first phase training data are used for building the model as well as estimating parameters using various classification techniques. Before training the model each attributes are normalized to zero mean and unit variances, which will improve performance of the model as well as cut down the learning time. Table 4 gives the variable of interest and the unit of measurement used in this study.

Table 4: Variables encoding summary

Code	Variables	Measurement
Y	Adoption behavior	Y = 0 for Non-Adopters = 1 for Adopters
X_1	Age of the farmer	Number of years
X_2	Education of the farmer	Formal Years of Education
X_3	Household Size	Number of family members
X_4	Farm Size	Number of acres
X_5	Farming Experience	Number of years
X_6	Animal Husbandry	Number of farm animals
X_7	Media Exposure	Number of sources exposed frequently
X_8	Extension Visits	Number of Visits
X_9	Crop Diversification	Number of Crops Grown in that year
X_{10}	Income Status	In Rupees (Rs.)
X_{11}	Worth of Liquidating Assets	In Rupees (Rs.)
X_{12}	Crop Insurance by Government	In Rupees (Rs.)

Support Vector Machine with Sigmoid Kernel trick

SVMs are a classification algorithm that was originally developed for binary classification problems, but can be generalized to handle multiple classes. The model adopted a single optimization problem to find the optimal hyper-parameters, which are estimated using an efficient random grid search technique. Two different kernels, sigmoid and RBF, were applied to the data. While there are many traditional statistical classification procedures available, their usefulness depends on assumptions and circumstances.

Table 5 summarizes the specifications of the SVM model used in this study. The model was developed to classify a binary dependent variable using 12 predictors. Two types of SVM classification were considered, and classification type 1 was chosen with a capacity value of $C = 2.40$, which determines the tradeoff between the two boundaries in the model. The table also provides information on the number and types of support vectors used in the model, as well as the chosen kernel functions and their associated parameters.

Table 5: SVM model summary

Model Specifications	Value
Number of independents	12
SVM type	Classification type 1 (Capacity = 2.40)
Kernel type	Sigmoid (Gamma = 0.096, Coefficient = 0.06)
Number of SVs	97 (93 bounded)
Number of SVs (0)	48
Number of SVs (1)	49

In classification problems, only two hyperparameters are needed to be defined by user i.e., the trade-off between model capacity and training error represented by C (Capacity) and the kernel parameter γ (Gamma). These hyperparameters are directly coded with real values within a given search space to randomly generate M number of initial particles of set S. Search space of hyper-parameters and C and γ are respectively restricted to ranges of $[C_{min}, C_{max}]$ and $[\gamma_{min}, \gamma_{max}]$ are randomly generated. The entire operations has been system inbuilt algorithms in STATISTICA 8 version software. The current research randomly searched for the optimal value of the capacity parameter C and found it to be 2.40. Random grid search algorithm was used to find the optimal hyperparameters, where a range of values between minimum and maximum with particle incremental value as an interval was specified. Other optimization techniques such as particle swarm optimization, genetic algorithms, ant colony optimization techniques, and simulated annealing algorithms can also be used to find optimal hyper-parameters.

For instance, a large C forces the SVM classification algorithm to reduce the training errors, which in turn can be accomplished by increasing the machine capacity and as a consequence may reduce the margin. This is contrary to the main objective of margin maximization and also does not guarantee a good generalization performance of the classifier. Therefore, selection of optimal hyper-parameters is an important step in SVM modeling. Further, several kernel functions $k(x_i, x_j)$ are available in the literature, like polynomial function, sigmoid kernel, Gaussian kernel, and RBF.

The current research has taken up with Sigmoid kernel and having hyperparameter gamma = 0.096, which was achieved by selecting the parameters range $[\gamma_{min}, \gamma_{max}]$. The $\gamma_{min} = 0$ and $\gamma_{max} = 10$ with the incremental value = 0.2 by providing this information to the random grid search end up with the saturated result of gamma $\gamma = 0.096$ and coefficient value mainly depends on how exactly the position and orientation of hyper plane, which depends on the weights which assign to the input vectors. It's ranged from 0 to 3 with the incremental value 0.02 wind up with the saturated final result 0.06. Set the iteration number (t) from 1 to maximum number of iterations and evaluate inertia weight $w(t)$ generation by generation according to the model equation. The current research has taken $I = 1000$ iterations for tuning the model.

SVM learning is based on the principle of structural risk minimization, which aims to minimize both the generalization bound and the empirical error, unlike traditional models that only minimize the empirical error. This technique is independent of the dimensionality of feature space and obtains generalization by maximizing the margin, which minimizes the weight vector in a canonical framework. The training examples closest to the maximum margin hyperplane are called "support vectors" and are sparse, lying on the boundary and summarizing the information required to separate the data. The current research has 97 overall support vectors, with 93 bounded support vectors, and 48 support vectors in the nonadopters category and around 49 support vectors in the adopters category. The result suggests that there are enough vectors on the boundaries to effectively classify cases into their respective classes. The hyperplane is making use of these vectors for classification.

Table 6: Classification matrix

Sample Observed	Predicted		
	Adopters	**Nonadopters**	**Percent Correct**
Training Adopters	62	8	88.5%
Non Adopters	8	42	84.0%
Overall Percent			86.2%
Testing Adopters	17	1	94.4%
Nonadopters	1	11	91.6%
Overall Percent			93.3%

Table 6 shows that the cells on the diagonal of the cross classification of cases are correct predictions for each sample. The cells off the diagonal of the cross classification of cases are incorrect predictions of the cases used to create the model, 62 of the 70 farmers who previously adopted the drought coping strategies are classified correctly; 42 of the 50 nonadopters are classified correctly. Overall, 86.2% of the training cases are classified correctly, corresponding to the 13.8% incorrect shown in the model summary table. A better model should correctly identify a higher percentage of the cases.

Classifications based upon the cases used to make the model tend to be too "optimistic" in the sense that their classification rate is inflated. The testing sample helps to validate the model; here 93.3% of these cases were correctly classified by the model. This suggests that overall our model is in fact correct and efficient in prediction and classification.

Support Vector Machine with RBF Kernel Trick

Here, the result explains the detail about the model summary and specifications of the SVM model, including the number of support vectors and their types, the kernels and their parameters. The current research has experimented most commonly used RBF kernel function $k(x_i, x_j) = \exp\{-\|x - x_j\|^2/2\sigma^2\}$ to train nonlinear SVM model. It has only one hyperparameter that needs to be pre-determined and yields good performance under general conditions. Training nonlinear SVR with RBF kernel function requires optimization of two hyperparameters, namely, (i) Regularization parameter C, which balances the complexity and approximation accuracy of the model and (ii) Kernel bandwidth parameter σ, which represents variance of RBF kernel function. However, here we can notice that in above RBF kernel function can be equivalently represented as $k(x_i, x_j) = \exp\{-\gamma\|x - x_j\|^2\}$, where $\gamma = 0.5/\sigma^2$. This is necessary due to SVM package requirements, since it works with gamma values and not directly with the width σ. Here we adopt $k(x_i, x_j) = \exp\{-\gamma\|x - x_i\|^2\}$ form RBF kernel function, which require value to be estimated instead of σ.

Table 7: Model summary

Model Specifications	**Value**
Number of independents	12
SVM type	Classification type 1 (Capacity = 1.200)
Kernel type	Radial basis function (Gamma = 0.083)
Number of SVs	82 (74 bounded)
Number of SVs (0)	41
Number of SVs (1)	41

Table 7 explains the model summary, the fitted model has 12 predictors and used the classification type 1 by using capacity parameter C = 1.200, the capacity parameter was computed by random grid approach and the current research has taken up with RBF as kernel tricks and having hyper parameter gamma = 0.083, which has achieved by selecting the parameters range $[\gamma_{min}, \gamma_{max}]$. The $\gamma_{min} = 0$ and $\gamma_{max} = 10$ with the incremental value = 0.02. It has 82 overall support vectors and having 74 bounded support vectors and coming to categorywise, there are 41 support vectors are in the side of nonadopters category and around 49 support vectors are in the side of adopters category. The result indicates there is sufficient number of vectors is at boundaries which are making the hyper plane for effectively classifying the cases into respective classes.

Table 8: Classification matrix

Sample Observed	**Predicted**		
	Adopters	**Nonadopters**	**Percent Correct**
Training Adopters	65	5	92.83%
Nonadopters	6	44	88.0%
Overall Percent			90.42%
Testing Adopters	13	5	72.2%
Nonadopters	1	11	91.6%
Overall Percent			81.44%

Table 8 shows that the cells on the diagonal of the cross classification of cases are correct predictions for each sample. The cells off the diagonal of the cross classification of cases are incorrect predictions of the cases used to create the model, 65 of the 70 farmers who previously adopted the drought coping strategies are classified correctly; 44 of the 50 nonadopters are classified correctly. Overall, 90.42% of the training cases are classified correctly, corresponding to the 9.58% incorrect shown in the model summary table. A better model should correctly identify a higher percentage of the cases.

Table 9: Classification ability measures

SVM MLP	Training	Testing	SVM RBF	Training	Testing
Accuracy	86.22	93.33	Accuracy	90.42	81.44
Recall	0.8857	0.9444	Recall	0.9285	0.7222
Precision	0.8857	0.9444	Precision	0.9154	0.9285
F1	0.8857	0.9444	F1	0.9219	0.8125

Table 9 presents the classification measures computed for two different variants of the SVM model. Both models demonstrate good performance across all measures, with an accuracy rate of more than 70% in both cases. The Recall and Precision measures also perform well for both models, with scores above 0.50. However, the SVM MLP model outperforms the SVM RBF model for the testing set, as evidenced by its higher Recall and Precision scores. Additionally, the F1 Score is high for both models, indicating their potential classification ability. Further comparative analysis of model performance confirms that the SVM MLP model is superior to the SVM RBF model.

Conclusion

SVMs were applied to classify the adoption of new technologies among dairy farmers. Two SVM variants with different sigmoid, kernel tricks, and RBF were evaluated, and their optimal hyperparameters, gamma and capacity, were identified using a random grid search optimization approach. Specifically, the SVM with sigmoid kernel had an optimal gamma value of 0.096 and a capacity C value of 2.40, while the SVM with RBF kernel had an optimal gamma value of 0.083 and a capacity C value of 1.20.

The SVM with sigmoid kernel had 97 overall support vectors, of which 93 were bounded, and 48 support vectors were in the nonadopters category, and 49 were in the adopters category. The SVM with RBF kernel had 82 overall support vectors, of which 74 were bounded, and 41 support vectors were in the nonadopters category, and 41 were in the adopters category. Both models had good classification performance, with accuracy rates above 70%, recall and precision scores above 0.50, and F1 scores above 0.70.

The SVM with sigmoid kernel achieved an overall classification accuracy of 86.2% on the training data and 93.3% on the testing data. The SVM with RBF kernel achieved an overall classification accuracy of 90.42% on the training data and 81.44% on the testing data. The outcomes suggest that the SVM with sigmoid kernel performed slightly better than the SVM with RBF kernel in terms of classification accuracy on the testing data. In summary, the SVM models with sigmoid and RBF kernels are effective in classifying the adoption of new technologies among dairy farmers. The models can be useful for decision-making in agriculture policy and extension services to identify and target farmers who are more likely to adopt new technologies.

R Code for using Support Vector Regression (SVR) for Binary Classification of Drought

```
# Load necessary libraries
library(e1071)
library(caret)
# Load the data
data <- read.csv("drought_data.csv")
# Split data into training and testing sets
set.seed(123)
trainIndex <- createDataPartition(data$Class, p = 0.7, list = FALSE)
trainData <- data[trainIndex, ]
testData <- data[-trainIndex, ]
# Train the SVR model
model <- svm(Class ~ ., data = trainData, type = "C-classification", kernel = "radial")
# Make predictions on the test data
predictions <- predict(model, testData[,-1])
# Evaluate the model
confusionMatrix(predictions, testData$Class)
```

References

Bharadwaj, A., Dahiya, S., Jain, R. (2012): "Discretization based Support Vector Machine (D-SVM) for Classification of Agricultural Datasets." International Journal of Computer Applications 40, no. 1 (2012): 975–8887.

Crossa, J., Pérez-Rodríguez, P., Cuevas, J., Montesinos-López, O., Jarquín, D., de los Campos, G. "Genomic Selection in Plant Breeding: Methods, Models, and Perspectives." Trends Plant Sciences 22, (2017): 961–75. doi: 10.1016/j.tplants.2017.08.011.

Devadas, R., Denham, R.J., Pringle, M. "Support Vector Machine Classification of Object-Based Data for Crop Mapping, Using Multi-Temporal Landsat Imagery." International Archives of the Photogrammetry, Remote Sensing and Spatial Information Sciences (2012): XXXIX-B7.

González-Camacho, J. M., Ornella, L., Pérez-Rodríguez, P., Gianola, D., Dreisigacker, S., Crossa, J. (2018): "Applications of Machine Learning Methods to Genomic Selection in Breeding Wheat for Rust Resistance." Plant Genome 11, (2018): 1-15. doi: 10.3835/plantgenome2017.11.0104.

Guo, X., Meng, C. (2015): "Research on Support Vector Machine in Image Denoising." International Journal of Signal Processing, Image Processing and Pattern Recognition 8, no. 2 (2015): 19-28.

Halagundegowda G. R., Singh, A., Meenakshi, H. K. "Discriminant Analysis For Classification of Farmers Based on Adoption of Drought Coping Mechanisms." Agriculture Update 12, (2017): 2643-48.

Hanamaratti, N.G., Prashanthi, S.K., Salimath, P.M., Hanchinal, R.R., Mohankumar, H.D., Parameshwarappa, K.G., Raikar, S.D. "Traditional Land Races of Rice in Karnataka: Reservoir of valuable traits." Current Science 94, no. 2 (2008): 242-47.

Jeetendra Prakash Aryal, M.L. Jat, Tek Bahadur Sapkota, Dil Bahadur Rahut, Munmum Rai, Hanuman S. Jat, P.C. Sharma, Clare Stirling. "Learning Adaptation to Climate Change from Past Climate Extremes: Evidence from Recent Climate extremes in Haryana, India." International Journal of Climate Change Strategies and Management, 2019.

Jog, S.M., Dixit, M., Rajgopalan, A. Ranade, S.D. "Balanced Vs Imbalanced Training Data: Classifying Rapideye Data with Support Vector Machines." International Journal of Innovative Research in Electrical, Electronics, Instrumentation and Control Engineering, 4, no. 6 (2016):158-163.

Kaiyi, W., Chunjiang, Z., Tian, Z.F., Juan, W.X., Hong, C., Fang, Y.Y. (2010). "Application of a Support Vector Machine to HACCP in the Animal Meat Industry." New Zealand Journal of Agricultural Research 50, no. 5 (2010): 743-48.

Nagaratna Biradhar, Sridhar (2004). "A Study on Consequences of 2003 Drought in Karnataka with Reference to Livestock and Fodder." Journal of Human Ecology 26, no. 2 (2004): 123-30.

Robothama, H., Castillo, J., Bosch, P., Perez-Kallens, J. "A Comparison of Multi-Class Support Vector Machine and Classification Tree Methods for Hydro Acoustic Classification of Fsh-Schools in Chile." Fisheries Research 111, (2011): 170–76.

Sahu, S. K., Prasad, M.B.N.V., Tripathy, B. K. "A Support Vector Machine Binary Classification and Image Segmentation of Remote Sensing Data of Chilika Lagloon." International Journal of Research in Information Technology 3 no. 5 (2015): 191-204.

Surase, R., Kale, K. (2015). "Multiple Crop Classification Using Various Support Vector Machine Kernel Functions." "International Journal of Engineering Research and Applications 5, no. 1 (2015): 103-9.

Surase, R., Kale, K. (2015). "Performance Evaluation of Support Vector Machine and Maximum Likelihood Classifier for Multiple Crop Classification." International Journal of Remote Sensing & Geoscience 4, no. 1 (2015): 62-9.

Ustuner, M., Sanli, F. B., Abdikan, S. 2016. This paper discusses the classification of rapid eye data using SVM, with a comparison between balanced and imbalanced training data.

Vapnik, V., Golowich, S., Smola, A. 1997. This classic paper introduced the SVM method for function approximation, regression estimation, and signal processing.

Wu, L., Wen, Y. 2009. This paper discusses the recognition of weed/corn seedlings using SVM and texture features.

Yang, Z., Gub, X. S., Lianga, X. Y, Linga, L.C. 2011. This paper presents a genetic algorithm-least squares SVM-based predicting and optimizing model on carbon fiber composite integrated conductivity.

Yao, T., Cheng, J., Xu, B., Zhang, M., Hu, Y., Zhao, J., Dong, X. W. 2015. This paper discusses the use of SVM classification models for rational design of novel tetronic acid derivatives as potent insecticidal and acaricidal agents.

Yao, X.J., Panaye, A., Doucet, J.P., Chen, H.F., Zhang, R.S., Fan, B.T., Liu, M.C., Hu Z.D. 2005. This paper presents a comparative classification study of toxicity mechanisms using SVM and radial basis function neural networks.

Zhao, M., Fu, C., Ji, L., Tang, K., Zhou, M. 2011. This paper proposes a new approach for feature selection and parameter optimization for SVM based on genetic algorithm with feature chromosomes.

Zolfaghari, K., Shang, S., McNairn, H., Li, J., Homyouni, S. 2013. This paper discusses the use of SVM for agriculture land use mapping with SAR data, with preliminary results from Western Canada.

Section 5

Hybrid Time Series Modeling for Water Resource Management

16

Revolutionizing Rainfall Prediction Boosting Accuracy with a Hybrid Statistical and Deep Learning Approach

G Avinash[1], Veershetty[1], Harish Nayak G H[1], Vinay HT[2] and Moumitha Baishya[1]

[1]*The Graduate School, ICAR-Indian Agricultural Research Institute, Library Avenue, New Delhi, India – 110 012*

[2]*Uttar Banga Krishi Viswavidyalaya, Pundibari, Coochbehar-736 165*

Abstract

Rainfall forecasting is the most critical and challenging task because of its dependence on different climatic and weather parameters. Hence, robust and accurate rainfall forecasting models need to be created by applying Deep Learning (DL) approaches and hybrid models. Several automatic systems were created to predict weather, but it depends on the type of weather pattern, season and location, which leads in maximizing the processing time. Therefore, in this work, a simple classical time series Auto-Regressive Integrated Moving Average (ARIMA) model is hybridized with Long Short-Term Memory (LSTM) deep learning network. In this study, monthly rainfall data of Coastal Karnataka over the period of 115 years (i.e., from 1901-2015) is utilized as more rain occurs in the region. The gathered information is analysed using ARIMA, LSTM and ARIMA-LSTM models, which are capable of processing the time series data and predict the dependency between the data effectively. The most challenging phase of ARIMA model is failing to capture the non-linearity. For obtaining the non-linearity deep learning neural network i.e., LSTM was hybridized. Proposed hybrid model was compared with the ARIMA and LSTM. Results indicated that the prediction accuracy of the proposed hybrid model is superior compared to other two competitive models for predicting the rainfall in Coastal Karnataka.

Keywords: *ARIMA, ARIMA-LSTM, Coastal Karnataka, Deep Learning, Forecasting, LSTM and Rainfall.*

Introduction

Agriculture is backbone of India's economy and is largely dependent on the monsoon. Rainfall plays a dominant role in agricultural production and productivity, despite advancement in many technological fronts. Forecast of rainfall is essential for planning and management of water resources especially in an agriculture-based country like India. About 65 per cent of the total cultivated land in India is under the influence of the rainfed agriculture system. Especially, monthly and seasonal rainfall forecasts provide useful information for water resource management, agricultural planning and its associated crop insurance application (Garbrecht *et al.,* 2010).

The early prediction of rainfall is essential which help farmers to make decisions on crop production and harvesting, as well as help prevent flooding, protect human lives and resources. Rainfall forecasting is very challenging due to its uncertain nature and dynamically changing climate. It is an application of science and technology to predict precipitation in advance. It's always been a challenging task for meteorologists. Prediction of precipitation is categorized into short-range prediction and long-range prediction. Forecasting is done through the collection and analysis of weather and climate data. Avinash *et al.* (2022) studied temporal variation of land use pattern in Karnataka through indices and concluded that due to fluctuations in the rainfall the forest land is degrading year by year.

Nowadays, Artificial Intelligence (AI) techniques are booming in the market, which are being used for data analysis and prediction purposes in different sectors Kala *et al.* (2022). In various papers for rainfall prediction of different Data Mining and Machine Learning (ML) techniques have been used Kanchan, P. (2021). These techniques show better predictive accuracy. A Deep Learning (DL) approach is used in this study to analyse the rainfall data of the Coastal Karnataka Subdivision. DL is capable of handling a vast amount of data and is capable of handling complex problems. In this study, a Long Short-Term Memory (LSTM) technique has been used for monthly rainfall prediction of the Karnataka subdivision. A hybrid model *i.e.,* ARIMA-LSTM was developed to capture both linearity and non-linearity for rainfall prediction which helps to increase accuracy of the model compare to the conventional models for development of policy making by the government and other precautionary measures to avoid critical situations.

Study Area

The highest amount of rainfall in Karnataka occurs in Coastal subdivision of Karnataka. It comprises of three districts Udupi, Uttar Kannada, and Dakshina Kannada. In this subdivision, the average annual rainfall is about 3456 mm. Karnataka experiences all the four types of climates *viz.,* pre-monsoon, monsoon, post-monsoon, and winter. Karnataka also receives high amount of rainfall during monsoon season.

Material and Methods

Box-Jenkins Time Series Model for Forecasting

Box-Jenkins procedure is concerned with fitting a mixed Auto Regressive Integrated Moving Average (ARIMA) model to a given set of data. The main objective in fitting this ARIMA model is to identify the stochastic process of the time series and predict the future values accurately. Originally, ARIMA models have been studied extensively by George Box and Gwilym Jenkins and their names have been frequently used synonymously with general ARIMA process applied to time series analysis, forecasting and control. However, the optimal forecasts of the future value of a time-series are determined by the stochastic process for that series. A stochastic process is either stationary or non-stationary. The first thing to note is that most time series are non-stationary and the ARIMA model refers only to a stationary (Makridakis *et al.*, 1998).

1. Autoregressive Integrated Moving Average (ARIMA) Model: ARIMA (p, d, q)

A generalization of ARMA models which incorporates a wide class of non-stationary time-series is obtained by introducing the differencing into the model. The simplest example of a non-stationary process which reduces to a stationary one after differencing is random walk. A process $\{y_t\}$ is said to follow an Integrated ARMA model, denoted by ARIMA (p, d, q), if $\nabla^d y_t = (1-B)^d \varepsilon_t$ is ARMA (p, q).

The model is written as $\varphi(B)(1-B)^d y_t = \theta(B)\varepsilon_t$

where $\varepsilon_t \sim WN(0, \sigma^2)$, *WN* indicating White Noise. The integration parameter d is a nonnegative integer. When $d = 0$, ARIMA $(p, d, q) \equiv$ ARMA (p, q).

The main stages in setting up a Box-Jenkins forecasting model are the following:

1. Identification: ACF plot, PACF plot, ADF test, KPSS test and PP test

2. Estimating the parameters: Maximum Likelihood Estimation
3. Diagnostic checking: Residuals of ACF, PACF plot, AIC and BIC values
4. Forecasting: accuracy of forecasts was tested using RMSE and MAPE

2. Long-Short Term Memory

Long Short-Term Memory (LSTM) is a specific recurrent neural network (RNN) architecture. It was proposed (Hochreiter *et al.*, 1997). Unlike a traditional feed-forward neural network, it includes feedback connections. Furthermore, it can be utilized on single-point data and the sequence of data as well. The essential components of LSTM are an input gate, an output gate and a forget gate, and the LSTM network was developed to resolve the vanishing gradient problem while training the traditional RNNs. LSTM is a cell memory unit that means that LSTM can remove or add information to the cell state. LSTM has overcome the vanishing gradients and the exploding gradients problem that appeared in RNN through the units' specific internal structure built in the model. Nowadays, LSTM has been known as a powerful method capable of processing, classifying, and making predictions based on time series data.

There are three gate controls: Input gate (i_t), Output gate (o_t), and Forget gate (f_t) in LSTM cell. The structure of the LSTM unit is shown in Fig.1 and the main calculation process is described in Fig.2 as follows

$$i_t = \sigma(W_{ix}x_t + W_{im}m_{t-1} + W_{ic}C_{t-1} + b_i)$$

$$f_t = \sigma\left(W_{fx}x_t + W_{fm}m_{t-1} + W_{fc}C_{t-1} + b_f\right)$$

$$c_t = f_t \odot c_{t-1} + i_t \odot g\,(W_{cx}x_t + W_{cm}m_{t-1} + b_c)$$

$$o_t = \sigma\left(W_{ox}x_t + W_{om}m_{t-1} + W_{oc}C_{t-1} + b_o\right)$$

$$m_t = o_t \odot h\left(c_t\right)$$

$$y_t = \varphi\,(W_{ym}m_t + b_y)$$

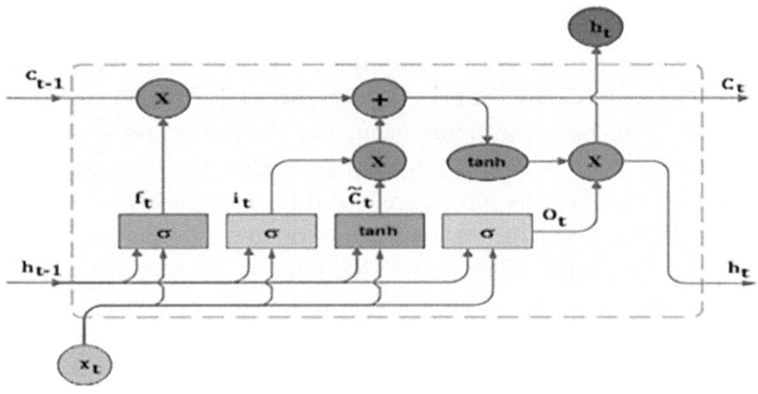

Fig. 1: Architecture of LSTM

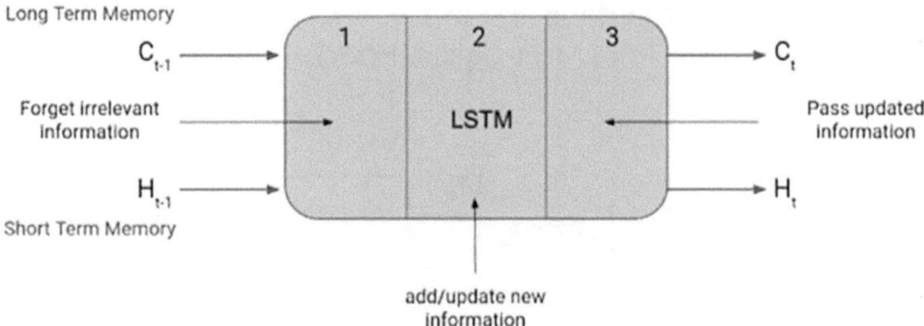

Fig. 2: Working of LSTM Neural Network

3. ARIMA-Long-Short Term Memory (ARIMA-LSTM) Hybrid Model

The hybrid method considers the time series Y_t as a combination of both linear and non-linear components. This approach follows the Zhang's (2003), Naveena *et al.* (2017a), Naveen *et al.* (2017b) developed hybrid approach with ARIMA-ANN model, accordingly the relationship between linear and nonlinear components can be expressed as follows

$$Y_t = L_t + N_t$$

Where, L_t and N_t represents the linear and nonlinear component respectively. In this work the linear part is modeled using ARIMA model and non-liner part by LSTM. The methodology consists of three steps. Firstly, an ARIMA model is employed to fit the linear component. Let the prediction series provided by ARIMA model denoted as $\hat{L}_t$. In the second step, the residuals ($e_t = Y_t - \hat{L}_t$) obtained from ARIMA model are tested for non-linearity by using BDS test (Brock 1996), once the residuals confirm the non-linearity, then they are modelled and predicted using LSTM. Finally, the forecasted linear and nonlinear components are combined to generate aggregate forecast as follows in the below equation.

$$\hat{Y}_t = \hat{L}_t + \hat{N}_t$$

Where, $\hat{L}_t$ and $\hat{N}_t$ represents the predicted linear and nonlinear component respectively. The graphical representation of hybrid methodology is expressed in Figure 3. Finally, the performance of the models under consideration are compared using RMSE (Root Mean Square Error).

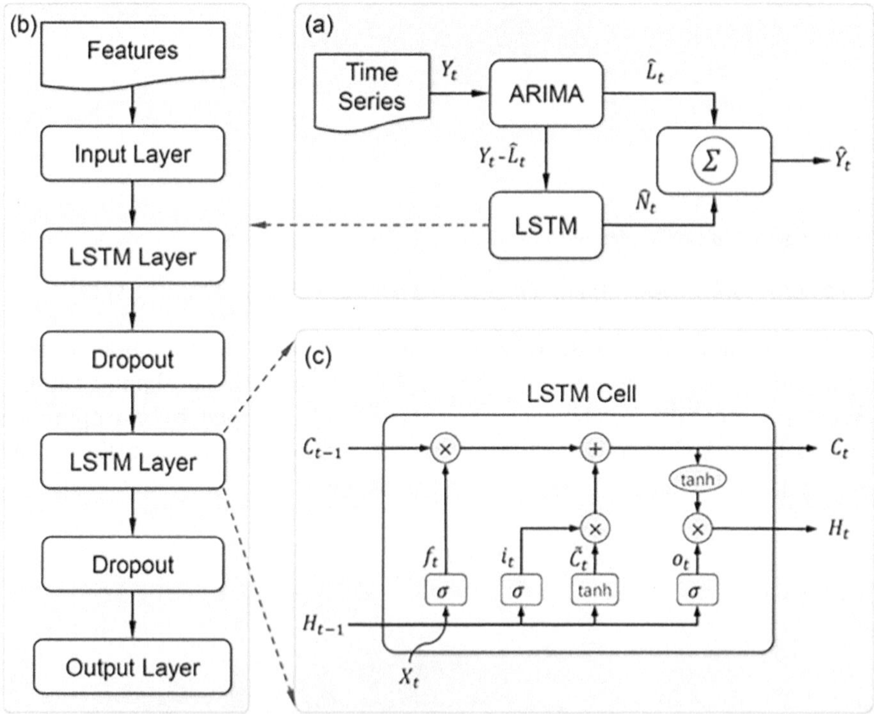

Fig. 3: The structure of the hybrid ARIMA-LSTM model for prediction of Rainfall
Source for Fig.1, Fig.2 and Fig.3 are from Datacamp website

Illustration

Forecasting of Monthly Rainfall of Coastal Karnataka Using Hybridization of Statistical and Deep Learning Technique

The forecasting of rainfall is an important part of the rainfall study as it helps in managing the rainfall issues. Heavy rainfall and low rainfall are two important changes in the rainfall pattern. The heavy rainfall, which leads to natural calamities like flood submergence of land under agriculture. Low rainfall results in difficulty to meet the water requirements. The proper study of rainfall issues may overcome these problems. Proper planning to avoid natural calamities like floods is very much essential. Managing the low rainfall through an alternative technique called irrigation and to conserve the water through water conservators *etc.*, requires study of rainfall for forecasting future scenario.

Monthly data on rainfall (in mm) of Coastal Karnataka was collected from the database "data.world" (https://data.world/rajanand/rainfall-in-india). For forecasting rainfall in Coastal Karnataka, data from 1901 to 2015 were obtained. The whole dataset was divided in to two parts, namely, training (1901-2014) and testing dataset (2015) were used for model building and to check the forecasting performance of the models, respectively. The ARIMA model has been built for the original time series which was found to be stationary series. The adequate model *i.e.*, ARIMA (2,0,1) has been identified based on Autocorrelation and Partial Autocorrelation Function (ACF and PACF) plots. The parameters of ARIMA models are estimated using maximum likelihood methods are given in Table 1. Further the model performance in testing data set is given in Table 4 along with accuracy measures.

Table 1: Parameter estimation of ARIMA (2,0,1) for rainfall time series data

Parameter	Estimate	Standard Error	t Value	Lag	p (residual) at first lag
Constant	284.03*	3.25	87.17	0	0.6291
AR1	1.45*	0.02	78.85	1	
AR2	-0.75*	0.02	42.04	2	
MA	-0.86*	0.01	72.98	1	

* Indicates significant at one percent level of significance

Most of the ML/DL models learn from the data by the time the learning model maps the data points from input to output. And the distribution of the data points can be different for every feature of the data. Larger differences between the data points of input variables increase the uncertainty in the results of the model. The ML/DL models provide weights to the input variables according to their data points and inferences for output. In that case, if the difference between the data points is so high, the model will need to provide the larger weight to the points and in final results, the model with a large weight value is often unstable. This means the model can produce poor results or can perform poorly during learning which causes overfitting and underfitting problem. So, scaling must be done to solve this problem.

$$\text{Scaling}=\frac{X_i - X_{min}}{X_{max} - X_{min}},$$

Where, X_{min}, X_{max} and X_i are the maximum, minimum and i[th] corresponding value for data taken.

One of the best deep learning techniques to capture long and short memory property is Long Short Term Memory (LSTM) was fitted to rainfall scaling dataset by fixing the number of batch size, activation functions, number of epochs, drop out layers to inactivate the neurons, loss functions and accuracy measures were considered. This methodology is called as trial and error method and its combinations were called as random search cross validation for tuning the parameter. Later, inverse scaling has to be done to obtain predicted values for comparison to the original data series. Further the model performance in testing dataset was given in Table 4.

Table 2: LSTM model specifications

Time series	Activation Function	Epochs	Loss function	LSTM nodes
Coastal Karnataka rainfall data	tanh	50	MSE	10

Table 3: Non linearity testing of ARIMA residuals by BDS test

Time series	Parameter	Dimension (m=2)		Dimension (m=3)	
Coastal Karnataka rainfall data	198.70	Statistic	probability	Statistic	probability
		31.38	<0.001	36.13	<0.001

After fitting ARIMA model, significance of residuals of ARIMA models were tested using BDS test. The BDS test in the above table shows that the residuals obtained from ARIMA models are nonlinear and significant. As, explained in hybrid methodology, once the residual series is found to be nonlinear, it can be modelled and predicted using nonlinear models. The nonlinear model like LSTM is used for modeling and forecasting of ARIMA residuals depicted in Table 4, in this study. After the confirmation of nonlinearity of ARIMA residuals, the same residuals are modelled and forecasted using LSTM model with the same specification given in Table 2. Further the forecasted residuals are combined with the forecasts obtained from original ARIMA model. Finally, the forecasting performance of ARIMA-LSTM hybrid model was fitted and forecasted one year (12 data points) data set is given in Table 4. The negative forecasted values is of no practical meaning and hence these values are set to zero for calculation of accuracy measures in various models.

Table 4: Comparison of forecasting performance of all models in test data set (2015)

Year-Month	Actual data	Forecast		
		ARIMA	LSTM	ARIMA-LSTM
2015-01	1.40	0.00	0.00	0.00
2015-02	1.00	33.81	0.00	0.00
2015-03	32.30	146.24	25.90	9.42
2015-04	72.20	271.03	25.78	0.00
2015-05	150.30	368.46	165.55	0.00
2015-06	735.30	416.78	297.73	439.59
2015-07	930.90	414.07	469.67	746.77
2015-08	575.20	373.87	1065.03	431.82
2015-09	260.30	317.34	722.04	162.33
2015-10	208.50	265.16	309.06	91.93
2015-11	124.20	231.55	187.83	304.53
2015-12	14.30	221.74	127.75	269.15
RMSE		218.19	271.84	**155.69**

As discussed in hybrid methodology section, the hybrid models have its own advantage over single models. Based on the lowest RMSE values of all models obtained for tested data set (Table 5) considered, one can infer that hybrid model consist of ARIMA and LSTM model outperformed over ARIMA and LSTM individually. Hybrid methodology considers both linearity and nonlinearity of the data set, hence, the performance of ARIMA-LSTM model is superior as compared to all other models under forecasting rainfall time series of Coastal Karnataka.

Conclusion

The primary conclusion of this study is that hybrid time series models perform better than single models. The hybrid model outperforms single time series in deep learning techniques for modelling and forecasting of rainfall time series data in Coastal Karnataka since the data set contained both linear and nonlinear patterns. The ARIMA with LSTM model, outperformed compare to other models in test data sets. In order to establish the model's practical validity, the hybrid approach can be further expanded utilising additional ML/DL approaches by varying autoregressive and moving average orders of linear time series models.

R codes

```
### ARIMA ###
# Load necessary libraries
library(forecast)
library(tseries)
library(ggplot2)
library(astsa)
library(lmtest)
library(tidyverse)
library(readxl)
library(writexl)
library(Metrics)
# Read the CSV file
data <- read.csv("C:\\Users\\haris\\OneDrive\\Desktop\\Rainfall\\Coastal_karnataka.csv",
header = TRUE)
attach(data)
# Display the head and tail of the data, and column names
head(data)
tail(data)
names(data)
# Plot the rainfall data
plot(data$rainfall, main = "Rainfall Data", xlab = "Time", ylab = "Rainfall")
# Divide data into training and test sets
# Convert the "rainfall" column to a time series
training_data <- ts(data$rainfall[1:1368], start = 1, frequency = 1)
test_data <- ts(data$rainfall[1369:1380], start = 1369, frequency = 1)
# Test autocorrelation in training data
Box.test(training_data, type = "Ljung-Box", lag = 1)
# Check for missing values and examine the summary of training data
sum(is.na(training_data))
summary(training_data)
class(training_data)
# Check stationarity of training data
```

```
autoplot(training_data, color = "black", main = "Time Series Plot") +
  xlab("Time") + ylab("Rainfall") +
  theme(plot.title = element_text(hjust = 0.5))
adf.test(training_data)  # H0: non-stationary
pp.test(training_data)  # H0: non-stationary
kpss.test(training_data) # H0: Stationary

acf2(ts(training_data), 100, main = "Autocorrelation Plot")
# Find the best ARIMA model
arimamodel <- auto.arima(training_data, ic = c("aicc", "aic", "bic"), trace = TRUE,
                    seasonal = FALSE, test = c("kpss", "adf", "pp"))
# Display ARIMA model information
summary(arimamodel)
coeftest(arimamodel)
arimamodel$fitted
# Define ARIMA model with specific order
ar201 <- Arima(training_data, order = c(2, 0, 1), seasonal = c(0, 0, 0), method = "ML")
# Display summary and check residuals
summary(ar201)
coeftest(ar201)
Box.test(ar201$residuals, type = "Ljung-Box")
bds.test(ar201$residuals, m = 2)
acf2(ar201$residuals, 40)
# Plot residuals and check for normality
plot.ts(ar201$residuals)
acf2(ar201$residuals, 20)
checkresiduals(ar201)
shapiro.test(ar201$residuals)  # H0: residuals are normally distributed
jarque.bera.test(ar201$residuals)
# Forecast for the next 12 observations
fcast1 <- forecast(ar201, h = 12)
# Display forecast plot and save the forecast to Excel
plot(fcast1)
write_xlsx(as.data.frame(fcast1),     "C:\\Users\\haris\\OneDrive\\Desktop\\Rainfall\\arima.
xlsx")
# Evaluate the performance of the ARIMA model
datatest <- as.numeric(test_data)
Arima_pred <- fcast1$mean
# Calculate metrics
MSE <- mse(datatest, Arima_pred)
RMSE <- sqrt(MSE)
MAPE <- mape(datatest, Arima_pred)
# Display metrics
RMSE
MAPE
###   ARIMA LSTM    ###
```

```
# After getting residuals from the ARIMA model,
# these residuals are used for capturing non-linearity.
# Min-Max normalization process
arima_residuals <- ar201$residuals # Replace with your actual ARIMA residuals
# Min-Max normalization process
max_residuals <- max(arima_residuals)
min_residuals <- min(arima_residuals)
# Scaling the residuals between zero and one
scaled_residuals <- as.data.frame(scale(arima_residuals, center = min_residuals, scale =
max_residuals - min_residuals))
# Load TSLSTM library
install.packages("TSLSTM")
library(TSLSTM)
# Build LSTM model
lstm_model <- ts.lstm(
  scaled_residuals,
  xreg = NULL,
  tsLag = 10,
  xregLag = 0,
  LSTMUnits = 10,
  DropoutRate = 0,
  Epochs = 50,
  CompLoss = "mse",
  CompMetrics = "mae",
  ActivationFn = "tanh",
  SplitRatio = 0.9910,
  ValidationSplit = 0.0087
)
# Display LSTM model details
lstm_model
# Obtain predicted values from the LSTM model
lstm_predicted <- as.data.frame(lstm_model$TestPredictedValue)
# Inverse scaling the predicted values
lstm_res <- lstm_predicted * (max_residuals - min_residuals) + min_residuals
# Save actual data to Excel
write_xlsx(data.frame(test_data), "H:\\R codes\\Arima_lstm_actual.xlsx")
# Save ARIMA forecast to Excel
write_xlsx(data.frame(fcast1$mean), "H:\\R codes\\Arima_test.xlsx")
# Save LSTM residuals to Excel
write_xlsx(data.frame(lstm_res), "C:\\Users\\haris\\OneDrive\\Desktop\\Rainfall\\lstmres.
xlsx")
# Adding forecasts of ARIMA and LSTM models
Arima_forecast <- data.frame(fcast1$mean)
LSTM_residuals <- data.frame(lstm_res)
Arima_lstm_pred <- Arima_forecast + LSTM_residuals
Arima_lstm_pred <- unlist(Arima_lstm_pred)
```

```
# Convert datatest to numeric
datatest <- as.numeric(datatest)
# Calculate metrics
library(Metrics)
mape_value <- mape(datatest, Arima_lstm_pred)
rmse_value <- rmse(datatest, Arima_lstm_pred)
# Display metrics
mape_value
rmse_value
###  LSTM  ###
library(TSLSTM)
library(tensorflow)
library(keras)
library(writexl)
attach(data)
names(data)
#Min-Max normalization process
max = max(data$rainfall)
min = min(data$rainfall)
#Scaling the residuals between zero and one
scaled_data = as.data.frame(scale(data$rainfall, center = min, scale = max - min))
scaled_data
lstm=ts.lstm(
  scaled_data,
  xreg = NULL,
  tsLag=10,
  xregLag = 0,
  LSTMUnits=10,
  DropoutRate = 0,
  Epochs = 50,
  CompLoss = "mse",
  CompMetrics = "mae",
  ActivationFn = "tanh",
  SplitRatio = 0.9910,
  ValidationSplit = 0.0087
)
lstm
a=lstm$TestPredictedValue
a=as.data.frame(a)
max=max(a)
min=min(a)
## Inverse scaling the predicted values
lstm_pred<- a*(max(data$rainfall)-min(data$rainfall))+min(data$rainfall)
model1=data.frame(test_data)
write_xlsx(model,"H:\\R codes\\lstm_actual.xlsx")
model2=data.frame(lstm_pred)
```

```
write_xlsx(model2,"C:\\Users\\haris\\OneDrive\\Desktop\\Rainfall\\lstmpred.xlsx")
# Adding forecasts of arima and lstm models
lstm_pred=unlist(model2)
datatest=unlist(model1)
datatest=as.numeric(datatest)
lstm_pred=as.numeric(lstm_pred)
library(Metrics)
mape=mape(datatest ,lstm_pred)
mape
rmse=rmse(datatest ,lstm_pred)
rmse
model=data.frame(lstm_pred)
write_xlsx(model,"H:\\R codes\\lstm_pred.xlsx")
############### END ##########################
```

References

Avinash, G., Ramesh, A. G., Rebasiddanavar, R. M., Shilpa, M. and Vinay, H. T. (2022). A study on temporal variation of land use pattern in Karnataka through indices.

Box, G.E.P. and Jenkins, G. (1970). Time series analysis, Forecasting and control, Holden-Day, San Francisco, CA.

Garbrecht, J. D., Zhang, X. C., Schneider, J. M., & Steiner, J. L. (2010). Utility of seasonal climate forecasts in management of winter-wheat grazing. Applied Engineering in Agriculture, 26(5), 855-866.

Hochreiter, S. and Schmidhuber, J. (1997). Long short-term memory. Neural computation, 9(8), 1735-1780.

Kala, A., & Vaidyanathan, S. G. (2022). Forecasting Monthly Rainfall using Bio-Inspired Artificial Algae Deep Learning Network. Fluctuation and Noise Letters, 21(02), 2250018.

Kanchan, P. (2021). Rainfall Analysis and Forecasting Using Deep Learning Technique. Journal of Informatics Electrical and Electronics Engineering (JIEEE).

Li, H., Qin, C., He, W., Sun, F., & Du, P. (2021). Improved predictive performance of cyanobacterial blooms using a hybrid statistical and deep-learning method. Environmental Research Letters, 16(12), 124045.

Makridakis, S., Steven, C. W. and Hyndman, R. J., 1998, Forecasting methods and applications, 3rd edition, John Wiley & sons, New York.

Naveena, K., Singh, S., Rathod, S., and Singh, A. (2017a). Hybrid ARIMA-ANN Modelling for Forecasting the Price of Robusta Coffee in India. International Journal of Current Microbiology and Applied Sciences, 6(7): 1721-1726.

Naveena, K., Singh, S., Rathod, S., and Singh, A. (2017b). Hybrid Time Series Modelling for Forecasting the Price of Washed Coffee (Arabica Plantation Coffee) in India. International Journal of Agriculture Sciences, 9(10): 4004-4007.

Zhang, G.P. (2003). Time series forecasting using a hybrid ARIMA and neural network model. Neurocomputing, 50, 159-175.

17

Leveraging the Potential of Artificial Intelligence Techniques for Time Series Analysis: A Case Study of Modeling Water Stress

Santosha Rathod[1], Amuktamalyada Gorlapalli[2] and Naveena K[3]

[1]ICAR-Indian Institute of Rice Research, India

[2]Professor Jayashankar Telangana State Agricultural University, India

[3]Centre for Water Resources Development and Management, Kozhikode

Abstract

Autoregressive Integrated Moving Average (ARIMA) models have a significant limitation in their assumption of linearity, which can lead to their inability to recognize nonlinear patterns present in time series data. Consequently, when time series contain nonlinear components, ARIMA models may be insufficient for accurate modeling and forecasting. To address this limitation, parametric nonlinear models can be used to capture nonlinear components, but they may still struggle to model highly complex, heterogeneous, and nonlinear data-generating processes. In contrast, artificial intelligence techniques such as Artificial Neural Networks (ANNs) and Support Vector Machines (SVMs) are more effective in capturing these types of patterns and are often preferred for modeling and forecasting time series data. This was demonstrated in the case of drought modeling in the Hyderabad region of Telangana, where Artificial Intelligence (AI) techniques outperformed classical ARIMA models in forecasting accuracy.

Keywords: *ANN, AI, drought modeling, hybrid models, SVR*

Introduction

Artificial intelligence (AI) is a field that mainly focuses on creating intelligent technologies that can easily perform tasks requiring human intelligence. One subfield is machine learning (ML), which enables machines to learn from data without being explicitly programmed. There are two main types of ML: supervised and unsupervised learning. Supervised learning trains machines on labeled data to predict output, while unsupervised learning trains machines on unlabeled data to find patterns. ML includes techniques such as regression, classification, clustering, and deep learning. ML has the potential to revolutionize many fields but also raises ethical, social, and legal issues. Developing trustworthy, responsible, and ethical ML systems is crucial. The MuCulloch and Pitts Model, proposed by Warren MuCulloch and Walter Pitts in 1943, is the first formal model of ML techniques. It is also known as the linear threshold gate (Fig. 1).

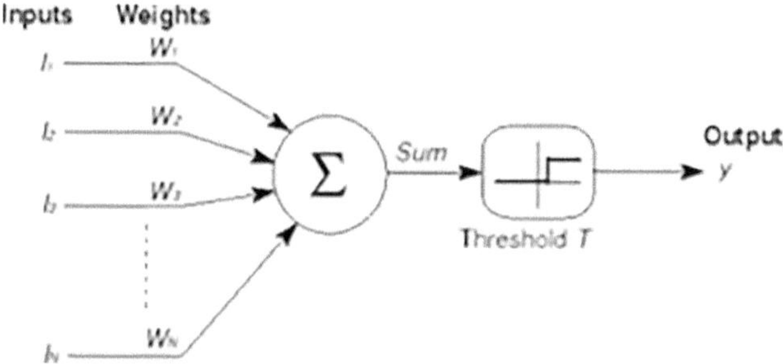

Fig. 1: Linear threshold gate.

The MuCulloch and Pitts Model is divided into two parts. The first part, g, performs an aggregation on the inputs, and the second part, f, makes a decision based on the accumulated value. The aggregation function g simply sums up the inputs, and the threshold parameter theta is set to a value such as 2. If the sum is 2 or greater, the decision function f outputs 0 (will watch the match), otherwise, it outputs 1 (won't watch the match). This is an example of threshold logic.

$$g(x1, x2, x3, \ldots\ldots xn) = g(x) = \sum_{i=1}^{n} xi$$

$$y = f(g(x)) = 1 \quad if\; g(x) \geq 0$$

$$= 0 \quad if\; g(x) < 0 \tag{1}$$

Frank Rosenblatt introduced the perceptron model in the late 1950s. It was a network of units that were an extension of the McCulloch-Pitts Threshold Logic Unit (TLU) model. The perceptron model included an additional input that represented bias.

$$sum = \sum_{i=1}^{n} Xi\ Wi + b \quad (2)$$

The McCulloch-Pitts TLU model paved the way for the development of neural networks, which have become one of the most promising and robust AI/ML techniques used in almost all areas of research and industry.

Time series analysis (TSA) is a statistical technique used to study trends and characteristics of data points collected in chronological order. A time series is an ordered sequence of values of a variable at equally spaced time intervals, and the goal of time series modeling is to develop a model that describes the inherent structure of the series, which can then be used to generate future values for the series, i.e., to make forecasts. Time series forecasting is crucial in various fields such as business, economics, finance, science, and engineering. The accuracy of forecasting models depends on the number of observations used in time series analysis, and it is generally believed that at least 50 observations are necessary to perform TSA. The Autoregressive Integrated Moving Average (ARIMA) model is one of the most widely used classical time series models due to its linear statistical properties and the general Box-Jenkins methodology for model building procedure.

The ARIMA model, along with other ML models, has been widely used in various fields for modeling and forecasting historical time series data. These applications include predicting coconut production in India (Naveena *et al.*, 2014), forecasting maize yield (Rathod *et al.*, 2017A), analyzing tomato supply and prices (Chitikela *et al.*, 2021), and predicting drought (Patil *et al.*, 2020; Gorlapalli *et al.*, 2022). Additionally, the ARIMA model has been utilized to forecast rice yield (Alam *et al.*, 2018B), predict coffee prices (Naveena *et al.*, 2017A; Naveena *et al.*, 2017B), estimate banana and mango yields (Rathod and Mishra, 2018B), analyze rainfall data (Saha *et al.*, 2020), and predict mango production time series (Rathod *et al.*, 2022). These applications demonstrate the adaptability and effectiveness of time series modeling in various domains.

Autoregressive Integrated Moving Average (ARIMA) Model

The ARIMA model is a widely used time series model due to its linear statistical properties and the popular Box-Jenkins methodology (Box and Jenkins 1970) for model-building procedures. Since many time series are nonstationary in nature, the ARIMA model introduces the differencing term d to obtain a stationary time series. The general form of the ARMA model becomes ARIMA (*p*,*d*,*q*) after adding the differencing term. The process Y_t follows an integrated ARMA model if $\Delta Y_t = (1-B)^d \varepsilon_t$., where d is the differencing parameter and ε_t represents the white noise error term. The ARIMA model can be expressed as follows:

$$\varnothing(B)(1-B)^d Y_t = \theta(B)\varepsilon_t \tag{3}$$

The Box-Jenkins methodology for building an ARIMA model involves three important steps: identification, estimation, and diagnostic checking. In the identification step, we need to identify the appropriate values of p, d, and q to use in the ARIMA model. This step involves analyzing the autocorrelation and partial autocorrelation plots of the time series data to determine the appropriate values of *p* and *q*, respectively, and using the differencing parameter d to make the series stationary. Once the appropriate values of *p*, *d*, and *q* have been identified, we move on to the estimation step. Here, we estimate the parameters of the ARIMA model using maximum likelihood estimation or a similar method. This step involves fitting the model to the time series data and obtaining estimates for the model parameters.

Finally, in the diagnostic checking step, we check the adequacy of the fitted model. This involves analyzing the residuals of the model to ensure that they are white noise, which indicates that the model is able to capture the underlying structure of the time series. If the residuals do not exhibit white noise properties, we need to revise the model by adjusting the values of *p*, *d*, and q or using a different model altogether. By following these three steps, the Box-Jenkins methodology provides a systematic approach to building an ARIMA model that can be used for time series analysis and forecasting.

Artificial Neural Network (ANN) for Time Series

Neural networks have become a popular method for time series analysis, and the Time Delay Neural Network (TDNN) is a type of neural network architecture that can be used for this purpose (Jha and Sinha 2012). TDNN uses time lags as inputs, which are considered in the input layer of the neural network. The general expression for the final output Y_t of a multi-layer feed-forward TDNN is given by equation (4), which includes model parameters, connection weights, the number of input nodes (*p*) and hidden nodes (*q*), and the activation function (*g*). The architecture of the neural network is shown in Figure 2. TDNN provides a flexible and powerful method for modeling and predicting time series data.

$$Y_t = \alpha_0 + \sum_{j=1}^{q} \alpha_j g\left(\beta_{0j} + \sum_{i=1}^{p} \beta_{ij} Y_{t-p}\right) + \varepsilon_t \quad (4)$$

where, $\alpha_j\ (j = 0,1,2,\ldots,q)$ and $\beta_{ij}\ (i = 0,1,2,\ldots,p, j = 0,1,2,\ldots,q)$ are the model parameters, also called as the connection weights.

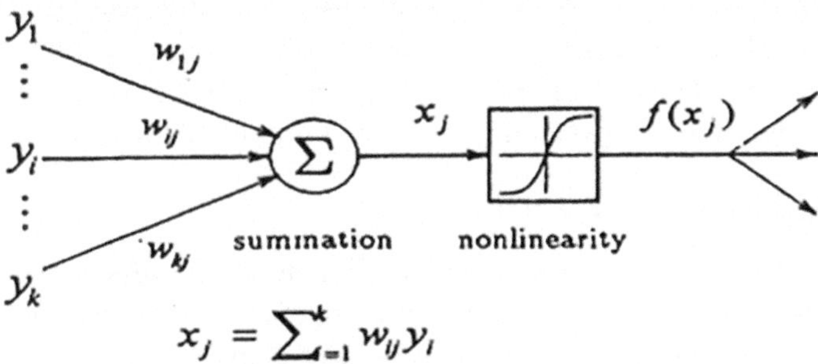

Fig. 2: Artificial Neural Network structure.

Support Vector Machine (SVM) for Time Series

SVM is a supervised ML technique, which was originally developed for linear classification problems. Later in the year 1997, the SVM for regression problems were developed by Vapnik by introducing ε-insensitive loss function (Vapnik, 1997) and it has been extended to the nonlinear regression estimation problems, and modeling of such problems is called as Nonlinear Support Vector Regression (NLSVR) model. The basic principle involved in NLSVR is to transform the original input time series into a high dimensional feature space and then build the regression model in a new feature space.

SVM is a powerful supervised ML technique that was initially developed for linear classification tasks. In 1997, Vapnik introduced the ε-insensitive loss function to extend SVM to regression problems. This development led to the creation of NLSVR models, which are used for nonlinear regression estimation problems (Vapnik, 1997). The NLSVR model works by transforming the original input time series into a high-dimensional feature space and then building the regression model in this new feature space. This process allows the NLSVR model to capture nonlinear relationships between input features and output targets. With the ability to handle nonlinear relationships, NLSVR is a valuable tool in time series analysis and forecasting. The general equation of NLSVR estimation function is:

$$f(x) = W^T \phi(x) + b \quad (5)$$

where, W is the weight vector, $\phi(x)$ is the feature vector, and b is the bias term. The function f(x) gives the predicted output for a given input x. The feature vector $\phi(x)$ is obtained by mapping the input vector x to a high-dimensional feature space using a nonlinear function.

The goal of NLSVR is to find the optimal Lagrange multipliers that minimize the empirical risk, subject to some constraints. The empirical risk is defined as the sum of the ε-insensitive loss function and a regularization term that controls the complexity of the model. The ε-insensitive loss function allows for a small deviation ε between the predicted output and the actual output.

The Hybrid Methodology

The hybrid method proposed in this work aims to model time series data as a combination of both linear and nonlinear components. This approach is based on the work of Zhang (2003) and involves three steps.

The first step is to model the linear component of the time series using an ARIMA model. This model is used to obtain a prediction series for the linear component, denoted as L^_t.

In the second step, the residuals ($e_t = y_t - \hat{L}_t$) obtained from the ARIMA model are tested for nonlinearity using the BDS test proposed by Brock (1996). If the residuals are found to be non-linear, they are modeled and predicted using either a TDNN or a NLSVR model.

Finally, the predicted linear and nonlinear components are combined to generate aggregate forecast.

$$\hat{y}_t = \hat{L}_t + \hat{N}_t \tag{6}$$

The predicted linear and nonlinear components, denoted as L̂ and N̂, respectively, are combined to generate an aggregate forecast. The performance of the models is evaluated using metrics such as Mean Square Error (MSE), Root Mean Square Error (RMSE), and Mean Absolute Percentage Error (MAPE).

The hybrid methodology presented in this study provides a promising approach for time series forecasting by combining the strengths of both linear and nonlinear models. To further improve the accuracy and practical validity of the hybrid model, other ML techniques can be incorporated for varying autoregressive and moving average orders. Additionally, the validity of the hybrid models can be further tested and generalized by applying this approach to a wider range of agricultural real-life datasets. This can provide a more comprehensive understanding of the effectiveness of the hybrid model in different contexts and help in identifying any limitations or areas for improvement.

Illustration

Water Stress Characterization and Prediction Using Artificial Intelligence and Time Series Models: A Study by Gorlapalli and Rathod et al (2022)

The study focuses on the Hyderabad region, which is the capital city of Telangana state in India. The area accepts most of its rainfall from the southwest monsoon, with an annual rainfall of approximately 750 mm. The researchers obtained monthly rainfall data (in mm) from January 1988 to May 2021 from the Directorate of Economics and Statistics, Government of Telangana. The data was then used to calculate the Standardized Precipitation Index (SPI) for three and six-month time scales. To predict drought at both SPI3 and SPI6 time scales, the study used AI models, namely ANN and Support Vector Regression (SVR). The methodology followed in the study is depicted in Figure 3.

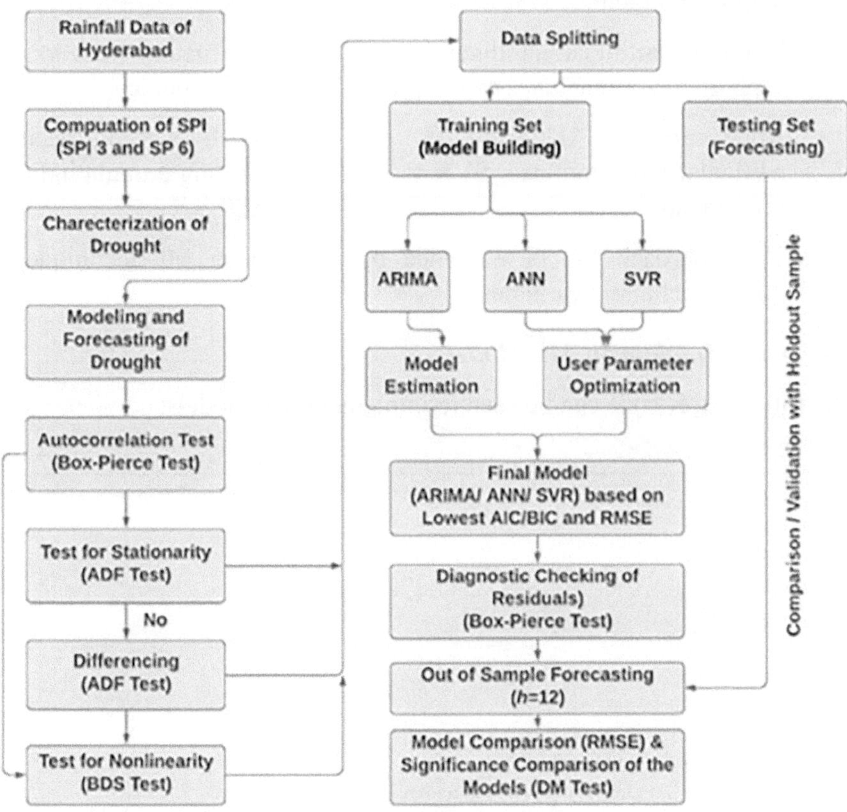

Fig. 3: Time Series and AI Models for drought prediction: Methodology overview.

Among the various ML models employed in the study, the SVR model demonstrated superior performance in both modeling and predicting drought conditions, as confirmed by the RMSE. The Diebold-Mariano test further substantiated the superior performance of the SVR model. The results indicated a significant reduction in the prediction error of SVR by 48% and 32% (compared to ARIMA), and by 21% and 26% (compared to ANN) for both SPI3 and SPI6 time scales in the test data sets.

The excellent performance of the SVR model may be attributed to its ability to effectively account for the nonlinear and complex patterns in the input data sets, unlike the classical linear ARIMA model. These results have significant implications for the sustainable and efficient management of water resources and stress in cropping systems.

Conclusion

- Classical time series models may not always be appropriate for data with nonlinear structures
- Nonlinear artificial intelligence techniques like neural networks and support vector machines can improve forecasting accuracy
- Application of AI techniques can increase forecasting accuracy, particularly the nonlinear SVR model for forecasting drought indices in Hyderabad
- This approach can be extended by using other ML techniques for different climatic variables

R Codes to Implement ML Models

The following R code can be used to implement ML models:

```
# Load required libraries
library(forecast)
library(e1071)
library(tseries)
library(ggplot2)
library(tidyverse)
library(fNonlinear)
library(lmtest)
# Read in the data
g=read.table(file="rf.txt",header=T)
# Perform Box-Ljung test for independence
Box.test(g$Rainfall)
```

```
# Read in and plot the data
rf1=read.table(file="rf1.txt",header=T)
head(rf1)
ggplot(data = rf1, aes(x = Month, y = Rainfall) )+
        geom_line(color = "#00AFBB", size = 1) +
        labs(x = "Months", y = "Rainfall") +
        ggtitle("TS Plot of Monthly Rainfall Data")
# Perform the BDS test for nonlinearity
bdsTest(g$Rainfall, m = 3, eps = NULL, title = NULL, description = NULL)
dim(g)
# Split the data into two parts for forecasting
a1=g$Rainfall[1:1416]
a2=g$Rainfall[1417:1428]
# Perform Box-Pierce test and plot ACF and PACF
Box.test(a1)
acf(a1)
pacf(a1)
The R codes implement the ARIMA model for time series analysis:
# Fit the ARIMA model and test for coefficients and accuracy
a1 <- g$Rainfall[1:1416]
m1 <- auto.arima(a1)
coeftest(m1)
accuracy(m1)
# Test for residuals
Box.test(m1$residuals)
# Generate fitted values and write to a file
fitted1 <- m1$fitted
write.csv(as.data.frame(fitted1), file="ARIMA_Fitted.csv")
# Generate forecasted values and calculate RMSE
a2 <- g$Rainfall[1417:1428]
f1 <- forecast(m1, h=12)
f12 <- f1$Point.Forecast
mse11 <- abs(a2-f12)^2
mse1 <- mean(mse11)
rmse1 <- sqrt(mse1)
# Write forecasted values to a file
write.csv(as.data.frame(f12), file="ARIMA_Forecasted.csv")
################### ANN ##########
# Fit an ANN model
m2 <- nnetar(a1, P = 1, size = 10, repeats = 25, xreg = NULL, lambda = NULL, model =
NULL, subset = NULL, scale.inputs = TRUE, maxit = 150)
# Generate forecasts using the ANN model
f2 <- forecast(m2, h = 12)
f22 <- f2$mean
mse21 <- (a2 - f22)^2
mse2 <- mean(mse21)
rmse2 <- sqrt(mse2)
```

```
write.csv(as.data.frame(f22), file = "ANN_Forecasted.csv")
# Fit another ANN model with default parameters
m3 <- nnetar(a1)
# Generate forecasts using the second ANN model
f3 <- forecast(m3, h = 12)
f32 <- f3$mean
mse31 <- (a2 - f32)^2
mse3 <- mean(mse31)
rmse3 <- sqrt(mse3)
write.csv(as.data.frame(f32), file = "ANN_Forecasted.csv")
################### SVR ##########
X1=g$Rainfall[1:1416]
Y1=g$Rainfall[2:1417]
X2=g$Rainfall[1416:1427]
Y2=g$Rainfall[1417:1428]
m4=svm(X1,Y1,degree = 3,cost = 45.69, nu=0.5,tolerance = 0.00001,epsilon = 0.00001)
summary(m4)
fitted4 <- fitted(m4)  ## Fitted values
mse41=abs(Y1-fitted4)^2
mse4=mean(mse41)
rmse4=sqrt(mse4)
rmse4
Box.test(m4$residuals)
s3=predict(model,X2)
mse61=abs(Y2-s3)^2
mse6=mean(mse61)
rmse6=sqrt(mse6)
############# ARIMA ###########
##########Significance Comparison ##########
########## For testing set ######
dm.test(m1$residuals, m2$residuals)
dm.test(m1$residuals, m3$residuals)
dm.test(m1$residuals, m4$residuals)
######## You have to do it for testing set also #####

########### Hybrid Modeling ##########
r1=m1$residuals
bdsTest(r1, m = 3, eps = NULL, title = NULL, description = NULL)
n1=nnetar(r1)
n1f=n1$fitted
c1=(m1$fitted)+n1f
c11=c1[32:1416]
a11=a1[32:1416]
mse51=abs(a11-c11)^2
mse5=mean(mse51)
```

```
rmse5=sqrt(mse5)
############ Comparison##########
accuracy(m1)
accuracy(m2)
################## Fitted Plots #########
rm(list=ls())
library(tidyverse)
library(readxl)
library(ggplot2)
Data1<-as.data.frame(read_excel("Fitted_Plot.xlsx", col_names = TRUE,sheet = "data"))
Date <- seq(as.Date("2020/1/06"), as.Date("2020/06/30"), "day")
RF=Data1$RF
Actual=Data1$Actual
Model1=Data1$Model1
Model2=Data1$Model2
Model3=Data1$Model3
Data2=data.frame(Date, RF, Actual,Model1, Model2, Model3)
df <- Data2 %>%
      select(Date, Actual, Model1, Model2, Model3) %>%
      gather(key = "Models", value = "RF", -Date)
p1<-ggplot(df, aes(x = Date, y = RF)) +
      geom_line(aes(color=Models), size=1)+scale_x_date(date_labels="%d/%b-%Y")+
      labs(x = "Date", y = "RF")+ ggtitle("Actual v/s Fitted plot RF")+
      theme(plot.title = element_text(size = 11))
p1+geom_vline(xintercept = as.Date("2020-06-24"), color="blue4")
```

References

Alam, W., Ray, M., Kumar, R.R., Sinha, K. S., Rathod, S., Singh, K.N. "Improved ARIMAX Modal Based on ANN and SVM Approaches For Forecasting Rice Yield Using Weather Variables." The Indian Journal of Agricultural Sciences 88, (2018): 1909–13.

Box, G.E.P., Jenkins, G. 1970. Time Series Analysis, Forecasting and Control. San Francisco, CA: Holden-Day.

Box, G.E.P., Jenkins, G.M., Reinsel, G.C. 1994. Time Series Analysis; Forecasting and Control. 3rd Edition, Englewood Cliff, New Jersey: Prentice Hall.

Brock, W.A., Dechert, W.D., Scheinkman, J.A, Lebaron, B. "A Test For Independence Based on The Correlation Dimension." Econometric Reviews, 15 :197–35.

Chitikela, G., Admala, M., Ramalingareddy, V.K., Bandumula, N., Ondrasek, G., Sundaram, R.M., Rathod, S. "Artificial-Intelligence-Based Time-Series Intervention Models to Assess the Impact of the COVID-19 Pandemic on Tomato Supply and Prices in Hyderabad, India." Agronomy 11, 2021: 1878.

Gorlapalli A., Kallakuri S., Sreekanth P.D., Patil, R., Bandumula, N., Ondrasek, G. "Characterization and Prediction of Water Stress Using Time Series and Artificial Intelligence Models." Sustainability 14, no. 11 (2022): 6690. https://doi.org/10.3390/su14116690.

Naveena, K., Rathod, S., Shukla, G., Yogish, K. J. "Forecasting of Coconut Production in India: A suitable Time Series model." International Journal of Agricultural Engineering 7, no. 1 (2014): 190–93.

Naveena, K., Singh, S., Rathod, S., Singh, A. "Hybrid ARIMA-ANN Modelling for Forecasting the Price of Robusta Coffee in India." International Journal of Current Microbiology and Applied Sciences 6, no. 7 (2017A): 1721–26.

Naveena, K., Singh, S., Rathod, S., Singh, A. "Hybrid Time Series Modelling for Forecasting the Price of Washed Coffee (Arabica Plantation Coffee) in India." International Journal of Agriculture Sciences 9, no. 10 (2017B): 4004–7.

Patil, R., Polisgowdar, B.S., Rathod, S., Satish, K.U., Srinivasa Reddy, G.V., Vijaya, W., Satyanarayana, R. "Drought Modelling and Forecasting using ARIMA and Neural Networks for Ballari District, Karnataka." Journal of the Indian Society of Agricultural Statistics 74, 2020: 149–15.

Rathod, S., Mishra, G.C. "Statistical Models for Forecasting Mango and Banana Yield of Karnataka, India." Journal of Agricultural Science and Technology 20, no. 4 (2018):803–16.

Rathod, S., Vijayakumar, S., Bandumula, N., Chitikela, G. "Prediction of Mango Production Using Machine Intelligence Techniques: A Case Study from Karnataka, India." Acta Scientific Agriculture 6, no. 9 (2022): 16–22.

Rathod, S., Mishra, G.C. "Statistical Models for Forecasting Mango and Banana Yield of Karnataka, India." Journal of Agricultural Science and Technology 20, 2018B: 803–16.

Rathod, S., Saha, A., Patil, R., Ondrasek, G., Gireesh, C., Anantha, M.S., et al. "Two-Stage Spatiotemporal Time Series Modelling Approach for Rice Yield Prediction & Advanced Agroecosystem Management." Agronomy 11, (2021): 2502. https://doi.org/10.3390/agronomy11122502

Rathod, S., Saha, A., Patil, R., Ondrasek, G., Gireesh, C., Anantha, M.S., et al. "Two-Stage Spatiotemporal Time Series Modelling Approach for Rice Yield Prediction & Advanced Agroecosystem Management." Agronomy 11, (2021): 2502. https://doi.org/10.3390/agronomy11122502

Rathod, S., Singh, K.N., Arya, P., Ray, M., Mukherjee, A., Sinha, K., Kumar, P., Shekhawat, R.S. "Forecasting Maize Yield Using ARIMA-Genetic Algorithm Approach." Journal Outlook Agriculture 46, (2017A): 265–71.

Rathod, S., Singh, K.N., Patil, S.G., Naik, R.H., Ray, M., Meena, V.S. "Modelling and Forecasting of Oilseed Production of India Through Artificial Intelligence Techniques." The Indian Journal of Agricultural Sciences 88, (2018A): 22–7.

Rathod, S., Yerram, S., Arya, P., Katti, G., Rani, J., Padmakumari, A.P., et al. "Climate-Based Modeling and Prediction of Rice Gall Midge Populations Using Count Time Series and Machine Learning Approaches." Agronomy 12, (2022): 22. https://doi.org/10.3390/agronomy12010022

Saha, A. Singh, K.N., Ray, M., Rathod, S. "A hybrid Spatio-Temporal Modelling: An Application to Space-Time Rainfall Forecasting." Theoretical and Applied Climatology 142, (2020): 1271–82.

Vapnik, V., Golowich, S., Smola, A. 1997. Support vector method for function approximation, regression estimation, and signal processing, In Mozer, M., Jordan, M and Petsche, T. (Eds). Advances in Neural Information Processing Systems. Cambridge, MA: MIT Press. 281–87.

Zhang, G.P. "Time Series Forecasting Using a Hybrid ARIMA and Neural Network Model." Neurocomputing 50, (2003): 159–75.

18

A Two-Stage Modeling Framework for Time-Series Analysis of Spatiotemporal Data

Santosha Rathod[1], Gayatri Chitikela[2], Amit Saha[3], Naveena K[4], Bishal Gurung[5], Mrinmoy Ray[5] and K N Singh[5]

[1]*ICAR-Indian Institute of Rice Research, India*

[2]*Professor Jayashankar Telangana State Agricultural University, India*

[3]*National Statistical Systems Training Academy, India*

[4]*Centre for Water Resources Development and Management, India*

[5]*ICAR-Indian Agricultural Statistics Research Institute, India*

Abstract

The Box-Jenkins Autoregressive Moving Average (ARMA) is the most popular model for predicting univariate time series data. However, the primary limitation of this model is that it considers only univariate series under consideration and ignores the neighborhood effects if any. Alternatively, the spatiotemporal auto-regressive moving average (STARMA) model is capable of modeling the observations recorded over both time and space. The spatial information in the model is incorporated using spatiotemporal weight matrices. The most commonly used spatial weight matrix is a uniform weight matrix in which all the neighbors are allocated equal weights, however, there have been many recent advances in the weighting schemes, which are offered based on the relatedness, heterogeneity, and distance among the neighbors. The weight matrices can be developed based on the scope and limitations of the modeling task under consideration. As STARMA is a linear type of spatiotemporal time series model that is incapable of capturing the nonlinear pattern present in spatiotemporal datasets, alternatively artificial Intelligence (AI) algorithms like artificial neural network (ANN) and support vector regression (SVR) are able to handle the nonlinearity component present in the data. In contrast, it is difficult to find complete linear or nonlinear data in the real world, although natural linear and nonlinear patterns are ubiquitous in spatiotemporal series. Neither the linear classical time series model nor the machine learning models are applicable under these conditions. In order to resolve this issue, two-stage modeling comes into the picture. The results demonstrated that a two-stage modeling framework outperformed single models for modeling and forecasting data sets.

Keywords: *ARMA, AI, spatiotemporal data, STARMA, two-stage modeling*

Introduction

The study of spatiotemporal time series forecasting has gained significant attention in recent years, as it involves analyzing records of events that occur over both space and time. Spatiotemporal models group time- and location-specific discrete variables and are used in various sectors, including agriculture, transportation, meteorology, economics, disease mapping, sociology, environmental sciences, and ecology. However, the popular Autoregressive Integrated Moving Average (ARIMA) (Box and Jenkins, 1970) model has two major flaws: it assumes linearity, which is not always the case in time series data that may exhibit nonlinear interactions, and it is a univariate model that cannot capture spatial linkages. Hence, when investigating time series data that are not perfectly linear and involve spatial correlation, the ARIMA model should not be used. Other multivariate time series models like VARMA have also been used to represent spatiotemporal data. However, VARMA models require more parameters, and the STARMA model introduced by Cliff and Ord has fewer parameters (Cliff and Ord, 1975). In previous literatures, the STARMA model was commonly used for modeling different spatiotemporal time series data (Kamarianakis, 2003). The STARMA model was previously employed to forecast the spatiotemporal data in different fields such as economics, disease mapping, climate modeling, agriculture, etc. (Pfeifer and Bodily, 1990; Lee, 2005; Rathod *et al.*, 2018; Saha *et al.*, 2020 and Rathod *et al.*, 2021).

Understanding the relationship between location and time can reveal important insights into the underlying causes of these phenomena (Schlenker and Roberts, 2009). Climate change, for example, can have far-reaching consequences across various domains, but detecting the tipping point at which these effects become significant requires the integration of time series and cross-sectional data (Martin Oeppen, 1975). In many cases, univariate statistical models fall short of capturing the complex spatial variations inherent in spatiotemporal data (Subba and Antunes, 2003). Therefore, modeling both temporal and spatial scales simultaneously can provide a more comprehensive understanding of inherent vulnerabilities across multiple sites during a particular period (Oud *et al.*, 2012).

Real-world spatiotemporal data often exhibits temporal, spatial, linear, and nonlinear dynamics, which are not fully captured by traditional time series models. The STARMA model is limited by its inability to handle nonlinear dynamics, which is often the case with real-world spatiotemporal data (Cheng *et al.*, 2014). While statistical models are available for handling nonlinearity, they often require assumptions that may not hold true in real-world scenarios. Alternatively, machine learning methods like artificial neural network (ANN) and support vector machine (SVM) have emerged as promising alternatives to model nonlinear patterns, due to their ability to capture the chaotic and complex relationship present in the data (Adhikari and Agrawal, 2014; Naveena *et al.*, 2017A; Naveena *et al.*, 2017B; Alam *et al.*, 2018A; Alam *et al.*, 2018B; Chitikela, *et al.*, 2020; Patil, *et al.*, 2020; Rathod *et al.*, 2021A; Rathod *et al.*, 2021B; Gorlapalli *et al.*, 2022). However, these models are generally suitable for univariate time series modeling (Rathod *et al.*, 2021). Hybrid techniques that combine statistical and machine learning models have been proposed to address this issue, but they have primarily focused on univariate time series data. In our previous studies, we have explored hybrid or two-stage modeling of spatiotemporal time series data (Saha *et al.*, 2022; Rathod *et al.*, 2021). In this chapter, effort has been made to explain the two-stage spatiotemporal modeling with one of our earlier published illustrated data (Saha *et al.*, 2020). This approach has significant implications for stakeholders involved in agro-food production, management, and planning.

STARMA Model

The STARMA model was initially proposed by Pfeifer and Deutsch (1980a) and later in a series of their publications, they delineated the parameter estimation procedure, variance co-variance structure, and sphericity test for residuals and correlated error structures (Pfeifer and Deutsch,1980b; Pfeifer and Deutsch, 1980c; Pfeifer and Deutsch, 1980d; Pfeifer and Deutsch, 1981). The space time model represents the dependence structure across both space and time. The dependent variable in the model is represented by $\boldsymbol{Z_i(t)}$, measured at fixed number of locations $I = 1, 2,\ldots,$ N at period T ($t = 1,2,\ldots,$ T) depicted in the form of autoregressive moving average form. The spatial locations can represent different geographic entities, such as a country, state, or region. The spatial weight matrix of size N × N represents the relationship between the neighborhood sites. The STARMA model can be expressed as a linear combination of past observations and errors, similar to a univariate time series.

Specifically, the STARMA ($p_{\lambda 1,}$ $\lambda_{2,}$ ..., λ_p, q_{m1}, m_2, ..., m_q) can be written as follows in the matrix equation:

$$Z(t) = \sum_{k=1}^{p} \sum_{l=0}^{\lambda_k} \phi_{kl} W^l Z(t-k) - \sum_{k=1}^{q} \sum_{l=0}^{mk} \theta_{kl} W^l \varepsilon(t-k) + \varepsilon(t) \quad (1)$$

where Z(t) = [$Z_1(t)$, ..., $Z_N(t)$] is a $N \times 1$ vector of observations at time t = 1, 2, ..., T; p is the autoregressive (AR) order; q is the moving average (MA) order; λ_k is the spatial order of the Kth AR term; m_k is the spatial order of kth MA term, ϕ_{kl} and θ_{kl} are the AR and MA parameters at temporal lag K and the spatial lag l (scalar); is the spatial weight matrix of size $N \times N$. The random error vector ε (t) = [ε_1 (t), ε_2 (t), ..., ε_N (t)]' is normally distributed at time t with E[ε(t)]

$$= 0, E\,[\varepsilon\,(t)\,\varepsilon\,'(t+s)] = \begin{cases} G = \sigma^2 I_N \;\; is\;\, s = 0 \\ 0,\; otherwise \end{cases} \text{ and } E[\varepsilon(t)\,\varepsilon'\,(t+s)] = 0, \text{ for } s > 0.$$

The STARMA model is again bifurcated as STAR and STMA models based on the presence and absence of AR and MA terms. In equation (1), when q = 0, only autoregressive terms remain model is named as STAR model and is be expressed as:

$$\boldsymbol{Z(t)} = \sum_{k=1}^{p} \sum_{l=0}^{\lambda_k} \phi_{kl} w^l Z(t-k) + \varepsilon(t) \quad (2)$$

when p = 0, resulting in the model becoming a space-time moving average (STMA) model, which is written as follows:

$$\boldsymbol{Z(t)} = \varepsilon(t) - \sum_{k=1}^{q} \sum_{l=0}^{m_k} \theta_{kl} w^l \varepsilon(t-k) \quad (3)$$

The STARMA model is constructed using the same three-stage technique for model creation as the Box-Jenkins univariate ARIMA technology, namely identification, estimate, and diagnostic testing as given by Pfeifer and Deutsch (1980b). The STARMA model becomes stationary if covariance matrix $Z(t)$ remains constant over a period of time and falls under unit root circle. The model parameter orders are identified based on space time autocorrelation function

(STACF) and the space time partial autocorrelation function (STPACF) by considering significant STAR and STMA spikes. The STACF and STPACF are depicted in equation number (4) and (5).

The STACF between l^{th} and k^{th} order neighbors separated by s time lags ($s = 1, 2, \ldots, k$ and h = 0, 1,.., λ) is presented:

$$\rho_{lk}(s) = \frac{\sum_{i=1}^{N}\sum_{t-1}^{T-S} W^{(l)} Z_i(t) W^{(k)} Z_i(t+s)}{\left[\sum_{i=1}^{N}\sum_{t=1}^{T-S} (W^{(l)} Z_i(t))^2 \cdot (W^{(k)} Z_i(t+s))^2\right]^{\frac{1}{2}}} \tag{4}$$

The STPACF is expressed as follows:

$$\rho_{h0}(s) = \sum_{j=1}^{k}\sum_{l=0}^{\lambda} \phi_{jl}\rho_{hl}(s-j) \tag{5}$$

Once the model orders are determined, then the model parameters are estimated using conditional least squares methods. If the variance-covariance matrix of the residuals is homogeneous, conditional least squares can be used to estimate the parameters. However, if the variance-covariance matrix is not homogeneous, conditional maximum likelihood estimation is used instead. Residual independence can be diagnosed using STACF and STPACF functions. If the residuals are approximately equivalent to white noise, then the sample STACFs should be practically zero.

Spatiotemporal Time Series Weight matrix

The spatiotemporal matrix is a crucial aspect of the STARMA modeling procedure as it captures the spatial relationships between the time series. However, determining the optimal number of spatial lag orders can be challenging as higher dimensions increase the computational complexity of parameter estimation. Therefore, the selection of spatial lag order involves a trade-off between prediction accuracy and computational feasibility. In most cases, a uniform and consistent spacing pattern is assumed for simplicity in model construction. The spatial weight matrix can be built in many ways depending on the light and scope the modeling procedure under consideration.

Uniform Weight Matrix

The standard STARMA models commonly employ a uniform weight matrix, also known as a homogeneous spatial weight matrix. However, this type of matrix is limited in that it does not consider spatial dynamics and heterogeneity.

As the number of spatial lags increases, the modeling and computational challenges also increase. To address these challenges and improve prediction accuracy, most research has focused on the second order spatial weight matrix (Sahu *et al*., 2019).

Euclidian Distance-Based Matrix

A Euclidean distance-based matrix has been developed as an alternative to the uniform weight matrix to account for spatial heterogeneity. This matrix is constructed using longitude and latitude of the sites and Euclidian distance calculated using the Riemannian great circle formula. The distance is calculated for each location, resulting in a matrix that reflects the spatial covariance structure. The formula used to calculate the Euclidean distance between the sites is

$$d_{ij} = r * a\cos\left\{\sin\left(\frac{lat_i}{57.2958}\right)\sin\left(\frac{lat_j}{57.2958}\right) + \cos\left(\frac{lat_i}{57.2958}\right)\cos\left(\frac{lat_j}{57.2958}\right)\cos\left(\frac{lon_j}{57.2958} - \frac{lon_i}{57.2958}\right)\right\} \quad (6)$$

where, $i, j = 1, 2, \ldots, N$, lat_i and lon_i are the latitude and longitude of site i, respectively and r is the radius of the earth. Building the first spatial order weight matrix takes into account inverse Euclidean distance (Subba and Antunes, 2003; Rathod *et al*., 2019). The calculated weights are then incorporated in the ARMA model to develop the heterogeneous spatiotemporal ARMA model.

K-Nearest Neighbor(KNN)-Based Correlation Weight Matrix

The KNN algorithm is a method for supervised learning that identifies the nearest neighbors based on distance or similarity matrices. In the context of STARMA modeling, this algorithm can be used to construct a spatial weight matrix by identifying the nearest neighbors of each location by KNN algorithm. Once the nearest neighbors have been identified, the linear Karl Pearson's correlation coefficient between different sites will be calculated and assigned as weights to the KNN sites. This approach can help to overcome the limitations of a uniform weight matrix by taking into account spatial heterogeneity and dynamics (Sahu *et al*,. 2019 and Rathod *et al*., 2021).

Fuzzy Rule-Based Spatiotemporal Weight Matrix

One significant drawback of employing uniform spatial weight matrices is their incapacity to account for spatial dynamics and heterogeneity adequately. To address this issue, we have created fuzzy inference system based spatiotemporal weight matrix in our previous work (Saha *et al.*, 2022). The fuzzy weights are calculated based on both the crow distance between the central points of the sites and the correlation of time series data. By using lower and upper limits for correlation and distance, a triangular fuzzy membership function is utilized to introduce fuzziness into the data. Selecting appropriate maximum and minimum weight restrictions helps in creating optimal fuzzy set (Saha *et al.*, 2019). Finally, the resulting matrix is normalized based on the defuzzified output.

Two-Stage Spatiotemporal Time Series Modeling

The two-stage methodology begins by examining the spatiotemporal relationship using a multivariate Box-Pierce test. Once the spatiotemporal relationship is established, the first stage involves developing the STARMA model on original data. After fitting the STARMA model, diagnostic checks are carried out. If the residuals are identified as white noise, the model can be considered finalized in the initial stage. However, if the residuals are found to be s significant and exhibit autocorrelation, then it is necessary to develop nonlinear or machine learning (ML) models on the residuals to correct the original STARMA model (Saha *et al.*, 2020; Rathod *et al.*, 2021). Since the data contain both linear and nonlinear patterns, neither STARMA nor nonlinear ML models are able to capture both linear and nonlinear spatiotemporal relationship.

$$Z_t = \hat{A}_t + B_t \tag{7}$$

where $\hat{A}_t$ stands for the value predicted using STARMA models, and B_t for the error term, which are nonlinear in nature and can be written as:

$$B_t = Z_t - \hat{A}_t \tag{8}$$

The residual B_t will then be predicted using ML models like ANN and SVR and fitted residuals by these models are then added to the fitted values of the STARMA models. Consequently, the combined forecast ultimate value can be expressed as follows:

$$\hat{Z}_t = \hat{A}_t - \hat{B}_t \tag{9}$$

The two-stage methodology is schematically depicted in Figure 1.

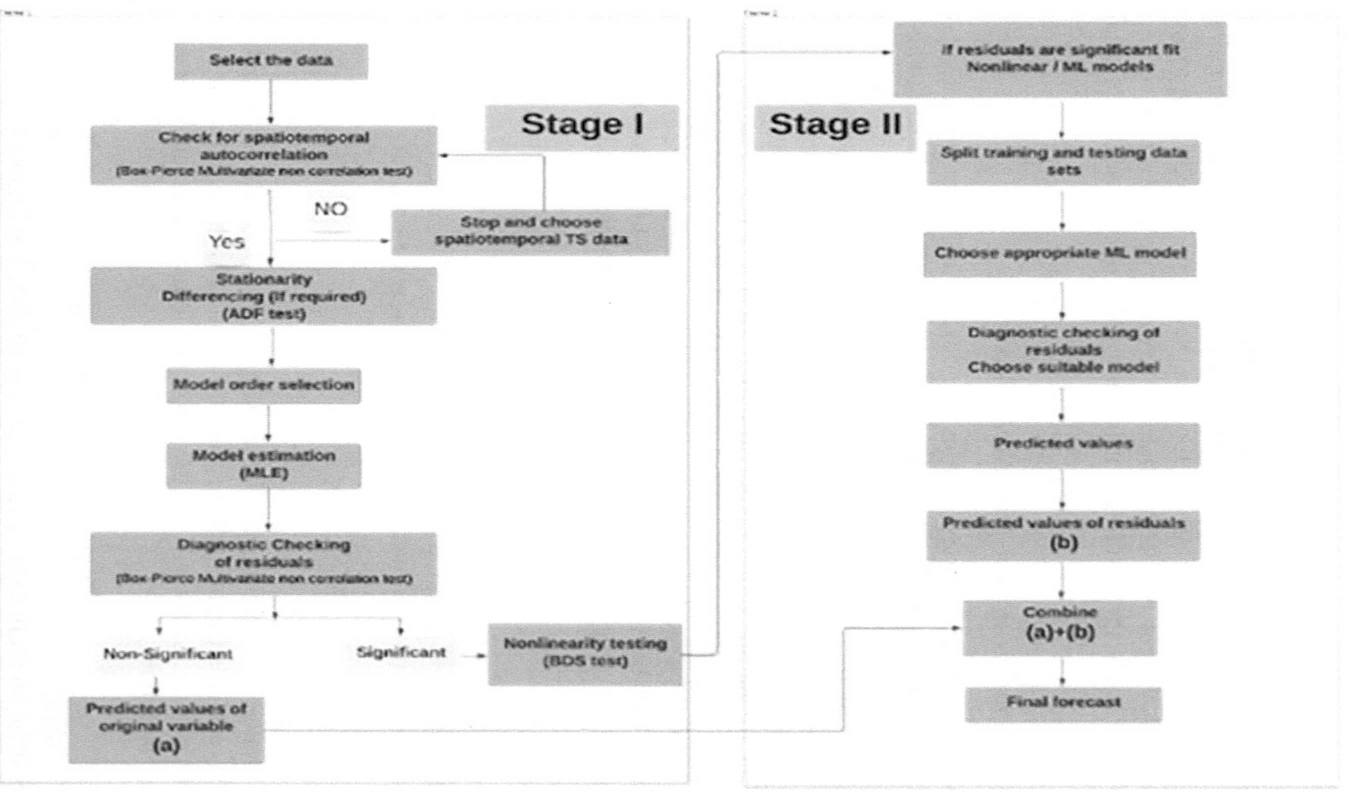

Fig 1: Schematic representation of methodology followed.

Illustration: Two-Stage Spatiotemporal Time Series Modeling of Annual Precipitation Data

We conducted a study (Saha *et al.*, 2020) on modeling and forecasting annual precipitation data of North Bengal. Data from six districts namely; South Dinajpur, Malda, Jalpaiguri, Cooch Behar, Darjeeling, and North Dinajpur were considered from the period of 1960 to 2015. The data set was divided into model building part (52 observations) and validation part (4 observations). Our proposed approach consists of two phases. In the first phase, we utilized a spatiotemporal autoregressive moving average (STARMA) model to capture the linear spatiotemporal patterns in the data. In the second phase, we employed SVR and ANN to capture the significant residuals (nonlinear part). Table 1 shows the comparison of our proposed two-stage modeling approach with other models in terms of mean absolute percentage error (MAPE) values.

Table 1: Model performance comparison in terms of Mean Absolute Percentage Error (MAPE)

Districts	ARIMA		STARMA		Two-Stage Approach	
	Training	**Testing**	**Training**	**Testing**	**Training**	**Testing**
South Dinajpur	14.04	7.39	16.21	19.03	2.94	3.64
Malda	13.14	11.55	17.41	32.80	6.80	10.62
Jalpaiguri	11.55	20.92	11.68	9.27	1.53	5.75
Cooch Behar	12.83	21.42	13.87	15.97	5.26	9.78
Darjeeling	19.9	11.71	10.89	4.49	2.43	4.74
North Dinajpur	12.59	15.59	13.96	38.34	2.28	8.17

Our results, as shown in Table 1, indicate that the proposed two-stage spatiotemporal hybrid approach outperformed both the STARMA and ARIMA models in terms of MAPE values for both the training and testing data sets. The MAPE values of proposed approach is less than the classical models. Table 2 provides the percentage increase in the prediction performance of our model in both the training and testing data sets.

Table 2: Proposed model performance (%) improvement in rainfall prediction accuracy compared to classical models

Data Splitting	STARMA over ARIMA	Hybrid Approach over STARMA	Hybrid Approach over ARIMA
Training	23.17	73.79	75.52
Testing	105.03	48.79	51.52

Our proposed model has demonstrated a significant improvement in prediction accuracy compared to classical models. Specifically, we observed a percentage increase in accuracy of over 70% in the training data set and over 45% in the testing data set. This improvement in performance can be attributed to our model's ability to consider both linear and nonlinear patterns in the data sets. Similar results have been reported in other studies, such as (Saha *et al.*, 2020; Rathod *et al.*, 2021), where two-stage models have also outperformed classical models. Overall, these findings suggest that our proposed approach is a promising alternative for modeling and forecasting spatiotemporal time series data which contains mixture of both linear and nonlinear spatiotemporal pattern in the time series. Further, Figure 2 clearly indicates fitted values of the proposed approach are closer to the actual precipitation data of six districts of North Bengal.

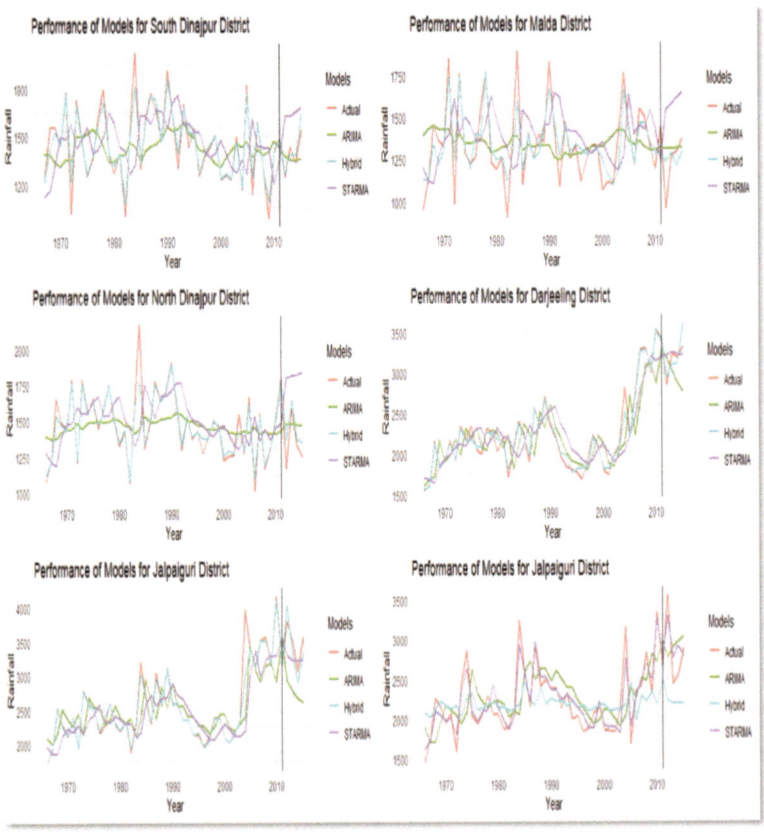

Fig 2: Observed vs fitted plot of precipitation data.

Conclusion

Spatiotemporal time series modeling helps to model the data recorded over both space and time in autoregressive and moving average lags. As STARMA is a linear model, it is unable to capture nonlinear spatiotemporal interactions or approximate complex nonlinear spatiotemporal patterns using linear approximations. On the other hand, ANN and SVR are promising machine learning algorithms for modelling the nonlinear patterns. However, neither STARMA nor ML models are able to model both linear and nonlinear components simultaneously. To address the challenge of mixed spatiotemporal linearity and nonlinearity, we have developed a two-stage modeling approach that effectively captures both linear and nonlinear aspects present in spatiotemporal time series data. For a contextual investigation, rainfall data North Bengal was considered to implement the proposed model. The planned two-stage methodology showed better prediction accuracy compared to classical models. In an illustrated data, the proposed approach exhibited prediction gains of over 75% in training data set and more than 50% in testing data sets respectively. The integration of the STARMA model with ML techniques offers a robust and effective approach for addressing the mixture of linear and nonlinear pattern present the spatiotemporal time series data. The proposed approach contributes to the developing theories in the area of spatiotemporal time series modeling and forecasting.

Code to Implement the Proposed Model In R Software

```
##Install the required package's ##

#install.packages("starma")
#install.packages("spdep")
#install.packages("forecast")
#install.packages("fNonlinear")
#install.packages("e1071")
#install.packages("lmtest")

##Load the required package's ##
library(starma)
library(spdep)
library("forecast")
library ("fNonlinear")
library("e1071")
library ("lmtest")

rm(list = ls())
## Import the input data ##
w0.mat=as.matrix(read.table(file="wo.txt",header=TRUE))
```

```
w1.mat=as.matrix(read.table(file="w1.txt",header=TRUE))
w0.mat
w1.mat
wlist =list(order0=w0.mat, order1=w1.mat)
wlist
st=as.matrix(read.table(file="ukavg.txt",header=TRUE)) # data read
stcor.test(st, wlist) #spatial corr
stacf(st, wlist, tlag.max=36)
stpacf(st, wlist, tlag.max=36)
#model fitting
#ar <- matrix(c(1, 1, 1, 0), 1,1)
#ma <- matrix(c(0, 1), 1, 2)
model=starma(st, wlist, ar = 1, ma = 1)
ab=summary(model)
capture.output(ab, file = "myfile.txt")
res=model$residuals
head(res)
fitted=st-res
fitted
fitted1=data.frame(fitted)
f1=fitted1$EngNE
f2=fitted1$EA
stdata=data.frame(st)
s1=stdata$EngNE
s2=stdata$EA
## Calculation of RMSE values
ms=(s1-f1)^2
ms1=mean(ms)*100
rmse1=sqrt(ms1)
rmse1
mss=(s2-f2)^2
ms2=mean(mss)*100
rmse2=sqrt(ms2)
rmse2
## Residual Diagnostic ##
r1=model$residuals
## Autocorrelation testing of residuals
Box.test(r1)
## Nonlinerity testing using BDS test
bdsTest(r1, m = 3, eps = NULL, title = NULL, description = NULL)
## ANN model fitting to STARMA residuals
## Hybrid modeling
n1=nnetar(r1)
n1f=n1$fitted
c1=(m1$fitted)+n1f
c11=c1[32:1416]
```

```
a11=a1[32:1416]
## RMSE Calculation
mse51=abs(a11-c11)^2
mse5=mean(mse51)
rmse5=sqrt(mse5)
##Significance Comparison of Model performance using DM test ##
dm.test(m1$residuals, m2$residuals)
dm.test(m1$residuals, m3$residuals)
dm.test(m1$residuals, m4$residuals)
```

References

Adhikari R., Agrawal R.K. "A Combination Of Artificial Neural Network and Random Walk Models for Financial Time Series Forecasting." Neural Computing & Applications 24, no. 6 (2014): 1441–49.

Alam, W., Ray, M., Kumar, R.R., Sinha, K.S., Singh, K.N. "Improved ARIMAX Modal Based on ANN And SVM Approaches for Forecasting Rice Yield Using Weather Variables." Indian Journal of Agricultural Sciences 88, (2018): 1909–13.

Alam W., Sinha K., Kumar R., Ray M., Rathod S., Singh K., Arya P. "Hybrid Linear Time Series Approach for Long Term Forecasting of Crop Yield." Indian Journal of Agricultural Sciences 88, (2018): 1275–79.

Box GEP, Jenkins G. 1970. Time series analysis. In: Forecasting and Control. San Francisco, CA: Holden-Day.

Cheng T., Wang, J., Haworth, J., Heydecker, B., Chow, A. "A Dynamic Spatial Weight Matrix and Localized Space-Time Autoregressive Integrated Moving Average for network modelling." Geographical Analysis. 46, (2014): 75–97.

Chitikela, G., Meena, A., Vijaya, K.R., Nirmala, B., Gabrijel, O., Raman, M.S., Rathod, S. "Artificial-Intelligence-Based Time-Series Intervention Models to Assess the Impact of the COVID-19 Pandemic on Tomato Supply and Prices in Hyderabad, India." Agronomy 11, no. 9 (2020): 1878. https://doi.org/10.3390/agronomy11091878

Cliff A.D., Ord, J. K. "Space-Time Modelling with an Application to Regional Forecasting." Transactions of Institute of British Geographers 64, (1975):119–128.

Deutsch, S.J. A three-stage iterative procedure for space-time modelling. Technometrics 22, no. 1 (1980a): 35–47.

Gorlapalli, A., Kallakuri S., Sreekanth P.D., Patil, R., Bandumula, N., Ondrasek, G., et al. "Characterization and Prediction of Water Stress Using Time Series and Artificial Intelligence Models." Sustainability 14, no. 11 (2022): 6690. https://doi.org/10.3390/su14116690

Kamarianakis Y. 2003. Spatial Time Series Modelling: A Review of the Proposed Methodologies. Working Paper, Regional Economic Applications Laboratory, Urbana, IL.

Lee C.Y. 2005. Space-Time Modelling and Application to Emerging Infectious Diseases, Doctorate Dissertation, Supervisor: B.K. Epperson, Michigan State University, Department of Forestry.

Martin R.L., Oeppen J.E. "The Identification of Regional Forecasting Models Using Space-Time Correlation Functions." Transactions of the Institute of British Geographers 66, (1975): 95–118.

Naveena, K., Singh, S., Rathod, S., Singh, A. "Hybrid ARIMA-ANN Modelling for Forecasting the Price of Robusta Coffee in India." International Journal of Current Microbiology and Applied Sciences 6, no. 7 (2017A): 1721–26.

Naveena, K., Singh, S., Rathod, S., Singh, A. "Hybrid Time Series Modelling for Forecasting the Price of Washed Coffee (Arabica Plantation Coffee) in India." Indian Journal of Agricultural Sciences 9, no. 10 (2017B): 4004–7.

Oud J. H. L., Folmer H., Patuelli, R., Nijkamp P. "Continuous-Time Modelling with Spatial Dependence." Geographical Analysis 44, (2012): 29-46.

Patil, R., Polisgowdar, B.S., Rathod, S., Satish, K.U., Srinivasa Reddy, G.V., Vijaya, W., Satyanarayana, R. "Drought Modelling and Forecasting using ARIMA and Neural Networks for Ballari District, Karnataka." Journal of the Indian Society of Agricultural Statistics 74, (2020): 149–15.

Pfeifer, P.E., Bodily, S.E. "A Test Of Space-Time ARMA Modelling and Forecasting with an Application to Real Estate Prices." International Journal of Forecasting 16, (1990): 255–72.

Pfeifer, P.E., Deutsch, S.J. "A comparison of Estimation procedures for the parameters of the STAR Model." B9, no. 3 1980: 255–70.

Pfeifer, P.E., Deutsch, S.J. "Independence and Sphericity Tests for the Residuals of Space Time ARIMA Models." Communication in Statistics, simulation and Computation B9, no. 5 (1980c): 533–49.

Pfeifer P.E., Deutsch, S.J. "Identification and Interpretation of First-order Space-time ARMA Models" Technometrics 22, no. 3 (1980b): 397–403.

Pfeifer P.E., Deutsch S.J. "Variance of the Sample-Time Autocorrelation Function of Contemporaneously Correlated Variables." SIAM of Applied Mathematics, series A. 40 no. 1 (1981): 133–36.

Rathod, S., Amit, S., Rahul, P., Gabrije, l. O., Channappa, G., Anantha, M.A., et al. "Two Stage Spatiotemporal Time Series Modelling Approach for Rice Yield Prediction & Advanced Agroecosystems Management. MDPI." Agronomy 11, (2021): 2502. https://doi.org/10.3390/agronomy11122502.

Rathod, S., Chitikela, G., Bandumula, N., Ondrasek, G., Ravichandran, S., Sundaram, R.M. "Modeling and Forecasting of Rice Prices in India during the COVID-19 Lockdown Using Machine Learning Approaches." Agronomy 12, no. 9 (2022): 2133. https://doi.org/10.3390/agronomy12092133

Rathod, S., Yerram, S., Arya, P., Katti G., Rani J, Padmakumari, A.P., et al. "Climate-Based Modeling and Prediction of Rice Gall Midge Populations Using Count Time Series and Machine Learning Approaches." Agronomy 12, no. 1 (2022): 22. https://doi.org/10.3390/agronomy12010022

Rathod, S. Gurung, B. Singh, K. N., Ray, M. "An Improved Space-Time Autoregressive Moving Average (STARMA) Model for Modelling and Forecasting of Spatio-temporal Time-Series Data." Journal of the Indian Society of Agricultural Statistics 72, (2018): 239–53.

Saha, A., Singh, K. N., Ray, M., Rathod, S. "A Hybrid Spatio-Temporal Modelling: An Application to Space-Time Rainfall Forecasting." Theoretical and Applied Climatology 142, (2020): 1271–82. https://link.springer.com/article/10.1007/s00704-020-03374-2

Saha, A., Singh, K.N., Ray, M. Rathod, S., Dhyani, M. "Fuzzy Rule–Based Weighted Space–Time Autoregressive Moving Average Models for Temperature Forecasting." Theoretical and Applied Climatolog. 150, (2022): 1321–35. https://doi.org/10.1007/s00704-022-04230-1

Saha, A. Singh, K.N., Ray, M. Kumar, S. Rathod, S. "A New Approach for Spatio-Temporal Modelling and Forecasting based on Fuzzy Techniques in Conjunction with K-Means Clustering." Journal of the Indian Society of Agricultural Statistics 73, (2019):111-20.

Sahu, P. K., Padhi, A., Patil, G.R., Mahesh, G., Sarkar, A. K. "Spatial Temporal Analysis of Freight Flow through Indian Major Seaport System." The Asian Journal of Shipping and Logistics 35, no. 9 (2019): 77–85.

Schlenker, W., Roberts, M. J. (2009). "Nonlinear Temperature Effects Indicate Severe Damages to U.S. Crop Yields Under Climate Change." Proceedings of the National Academy of Sciences 106, no. 37 (2009). 15594-98. DOI: 10.1073/pnas.0906865106.

Subba, R.T., Antunes, A. (2003). Spatio-Temporal Modelling of Temperature Time Series: A Comparative Study. In Time Series Analysis and Applications to Geophysical Systems. Schonberg, F.P., Brillinger, D. R., Robinson, E. A., (Eds.). New York, NY, USA: IMA Publications/Springer. pp. 123–50.

Index